Chemical Kinetics and Catalysis

FUNDAMENTAL AND APPLIED CATALYSIS

Series Editors: M. V. Twigg
Johnson Matthey
Catalytic Systems Division
Royston, Hertfordshire, United Kingdom

M. S. Spencer
School of Chemistry and Applied Chemistry
University of Wales College of Cardiff
Cardiff, United Kingdom

CATALYST CHARACTERIZATION: Physical Techniques for Solid Materials
Edited by Boris Imelik and Jacques C. Vedrine

CATALYTIC AMMONIA SYNTHESIS: Fundamentals and Practice
Edited by J. R. Jennings

CHEMICAL KINETICS AND CATALYSIS
R. A. van Santen and J. W. Niemantsverdriet

DYNAMIC PROCESSES ON SOLID SURFACES
Edited by Kenzi Tamaru

ELEMENTARY PHYSIOCHEMICAL PROCESSES ON SOLID SURFACES
V. P. Zhdanov

PRINCIPLES OF CATALYST DEVELOPMENT
James T. Richardson

A Continuation Order Plan is available for this series. A continuation order will bring delivery of each new volume immediately upon publication. Volumes are billed only upon actual shipment. For further information please contact the publisher.

Chemical Kinetics and Catalysis

R. A. van Santen and
J. W. Niemantsverdriet
Schuit Institute of Catalysis
Eindhoven University of Technology
Eindhoven, The Netherlands

Plenum Press • New York and London

Library of Congress Cataloging-in-Publication Data

On file

ISBN 0-306-45027-5

A Division of Plenum Publishing Corporation
233 Spring Street, New York, N. Y. 10013

10 9 8 7 6 5 4 3 2 1

Printed in the United States of America

FOREWORD
to the
Fundamental and Applied Catalysis Series

Catalysis is important academically and industrially. It plays an essential role in the manufacture of a wide range of products, from gasoline and plastics to fertilizers and herbicides, which would otherwise be unobtainable or prohibitively expensive. There are few chemical- or oil-based material items in modern society that do not depend in some way on a catalytic stage in their manufacture. Apart from manufacturing processes, catalysis is finding other important and ever-increasing uses; for example, successful applications of catalysis in the control of pollution and its use in environmental control are certain to increase in the future.

The commercial importance of catalysis and the diverse intellectual challenges of catalytic phenomena have stimulated study by a broad spectrum of scientists, including chemists, physicists, chemical engineers, and material scientists. Increasing research activity over the years has brought deeper levels of understanding, and these have been associated with a continually growing amount of published material. As recently as sixty years ago, Rideal and Taylor could still treat the subject comprehensively in a single volume, but by the 1950s Emmett required six volumes, and no conventional multivolume text could now cover the whole of catalysis in any depth. In view of this situation, we felt there was a need for a collection of monographs, each one of which would deal at an advanced level with a selected topic, so as to build a catalysis reference library. This is the aim of the present series, *Fundamental and Applied Catalysis*. Some books in the series deal with particular techniques used in the study of catalysts and catalysis: these cover the scientific basis of the technique, details of its practical applications, and examples of its usefulness. An industrial process or a class of catalysts forms the basis of other books, with information on the fundamental science of the topic, the use of the process or catalysts, and engineering aspects. Single topics in catalysis are also treated in the series, with books giving the theory of the underlying science, and relating it to catalytic practice. We believe that this approach provides a collection that is of value to both academic and industrial workers. The series editors welcome comments on the series and suggestions of topics for future volumes.

Martyn Twigg
Michael Spencer

Royston and Cardiff

PREFACE

The reaction rate expressions of a catalytic process provide the chemical engineer with essential information on the performance of a catalyst. The parameters in the kinetic expressions, such as the order of the reaction and the activation energy, are determined by the chemistry of the process. Understanding the full meaning of these parameters requires a thorough knowledge of the molecular basis of reaction kinetics. Insight at this level is essential for the development of new catalytic reactions or the improvement of existing catalytic technology.

During the past ten years it has become clear that there is not a large gap between industrial processes, which operate at high pressures, utilizing complex catalysts and high conversion rates, and laboratory experiments, which use pressures down to the ultrahigh vacuum range, with well-defined single-crystal surfaces as the catalysts and differential conversion levels. For several catalytic processes, such as the ammonia synthesis and the oxidation of carbon monoxide, it has been possible to predict kinetic behavior under process conditions from laboratory data obtained in surface science studies. Another important development in this field has been that people are beginning to understand the chemical kinetics of oscillating, exploding, and chaotic reactions.

The thermodynamics of reversible processes can be considered a scientifically mature field. Kinetics, however, which belongs to the discipline of irreversible thermodynamics, is still developing. The same is true for the dynamics of reacting molecules, which is a field of intensive research. As described in Chapter 1, the historical development of kinetics and catalysis as scientific disciplines started in the nineteenth century. Both fields experienced a major breakthrough in the beginning of this century.

The kineticist likes to reduce his problem so that optimum use can be made of thermodynamics by treating the difficult time-dependent part separately. This requires approximations which turn out to be very useful and widely applicable. Concepts such as the steady-state approximation, as well as sequences of elementary reaction steps in which one determines the rate of the entire sequence, are introduced in Chapter 2 and illustrated in Chapter 3.

In this book we intend to make a connection between molecular properties of species involved in catalytic reactions, their reactivity, and the expression for the reaction rate. Relations between the vibrational and rotational properties of molecules and their propensity to adsorb and react on a surface and to desorb into the gas phase, are derived from the field of statistical thermodynamics. The latter forms a substantial part of Chapter 4, where we also introduce the transition-state reaction-rate theory.

Although it is convenient to describe a reaction as if it occurs in isolation, interactions of the reacting species with other molecules are essential for letting the reaction event occur. Medium effects and energy exchange between reacting molecules and their surroundings are the subject of Chapter 5. The knowledge gained in the first five chapters is applied in Chapter 6, where we present an integrated microscopic description.

This book represents the third refinement of material that has been used over the past three years for the course "Chemical Kinetics and Catalysis" in the Schuit Institute of Catalysis at the Eindhoven University of Technology. We thank the students who participated in the course during the past three years; their comments, questions, and suggestions have played an important role in the revision of the previous versions. We are particularly grateful to Hannie Muijsers, Joop van Grondelle, Ton Janssens, and Tiny Verhoeven for their assistance in the production of text and figures.

Rutger van Santen
Hans Niemantsverdriet

Eindhoven

CONTENTS

CHAPTER 1. THE SCIENCE OF CATALYSIS

CHAPTER 2. THE RATE EQUATION

CHAPTER 3. INTRODUCTION TO CATALYTIC REACTIONS

CHAPTER 4. COLLISION AND REACTION-RATE THEORY

CHAPTER 5. MEDIUM EFFECTS ON REACTION RATES

CHAPTER 6. MICROSCOPIC THEORY OF HETEROGENEOUS CATALYSIS

THE SCIENCE OF CATALYSIS

1.1. INTRODUCTION

Catalysis and the growth of the chemical industry are closely connected. Heterogeneous catalysis enabled the development of large scale continuous processes.

This chapter presents an overview of the different aspects of catalysis as a science, and aims to provide a general background to the reader. In order to understand catalysis as it is today, it is important to have a sense of its historic origin, and of developments in related fields such as thermodynamics, kinetics, and chemical engineering. Although catalysis as a chemical reactivity phenomenon was well known in the nineteenth century, the field developed mainly in the twentieth century.

The expansion of the chemical process industry finds its roots in scientific innovations at the beginning of this century. These constitute the origin of industrial heterogeneous catalysis. Going from batch to continuous processes was the major process change that occurred. Continuous operation enables the efficient upscaling of conversion processes to very large scales, and is accompanied by immense economic advantages. Such processes were made possible by the introduction of heterogeneous catalysis in flow reactors, where products are continuously separated from the catalyst.

The century of catalysis that followed shows a beneficial symbiosis of industrial and scientific developments. It is a history of ongoing process innovations giving rise to new products and materials, with enormous benefits to our society. These developments were often driven by changing industrial needs or societal requirements, such as altering availability of raw materials, the need to reduce process constraints, or the demands for new products. Table 1.1 summarizes the most important changes that occurred in the chemical industry of the twentieth

TABLE 1.1. Major Phases in the History of the Chemical Industry

Before 1900	Heavy industry based on inorganic chemistry sulfuric acid, chlorine, soda
~1913	Introduction of continuous processes
1900–1940	Coal-based industry of base chemicals ammonia synthesis nitric acid fat hardening hydrodesulfurization/hydrodenitrogenation methanol synthesis
1940–now	Petrochemical industry fuels synthetic rubber polymers selective oxidation hydroformylation and other homogeneous reactions
~1970	New processing guided by efficient use of energy and raw materials environmental catalysis fine chemicals/homogeneous catalysis

century. Section 1.5 gives a more detailed overview of the evolution in industrial catalysis.

The two main developments in science that initiated industrial catalysis were the discoveries of catalytic hydrogenation by Paul Sabatier and the ammonia synthesis by Fritz Haber, which built upon the chemical equilibrium thermodynamics of Jacobus van't Hoff and the rate equation of Svante Arrhenius. Many processes, often based on catalytic hydrogenation, followed. Section 1.2 gives a historic review of the early developments in catalysis. Because of its pivotal role in the initiation of industrial catalysis, we devote an entire section to the development of the ammonia synthesis.

Catalysis as a science developed along three major directions: kinetics, catalyst characterization, and the synthesis of catalysts. Kinetics is the aspect of catalysis that we emphasize in this book. The basic kinetic characteristic of a catalytic reaction, as we recognize it now, is that it is a reaction cycle consisting of several coupled reactions. The kinetic formalism needed to treat coupled reaction systems dates from the first quarter of this century, as will be reviewed in Section 1.3.

A key development in reaction-rate theory was the introduction of activated complex or transition-state theory. This is a powerful formalism which enables one to predict the rate constant of a reaction step based on knowledge of the energetics and dynamics of the reactant molecules and their intermediates formed in the course of the reaction. Owing to the advance of modern spectroscopic and computational tools, direct information on the activated complexes and short-lived intermediates has become available, thus providing a molecular basis to kinetics. We will use transition-state theory as the basis of our treatment of the molecular basis of heterogeneous catalytic reactions.

In addition to the study of catalyst reactivity, characterization as well as synthesis of catalysts form the two other cornerstones of catalysis. We will briefly discuss the relation between the three main subdisciplines in the last section of this chapter.

1.2. THE EARLY DAYS OF CATALYSIS

Berzelius defined catalysis as a phenomenon that changes the composition of a reaction mixture but leaves the catalyst unaltered.

Catalysis started to be known in the first decades of the nineteenth century (see Tables 1.2 and 1.3). Although the earlier literature contains several reports of reactions that in hindsight must have been catalytic, these results played no role in the development of catalysis as a scientific discipline. In fact, the first catalytic process was already started around 1740. It was the Bell process for the production of sulfuric acid, in which SO_2 is oxidized by NO_2 to SO_3 and NO, while air regenerates NO to NO_2. Although the process represents an application of homogeneous catalysis, it is doubtful whether anyone appreciated the catalytic nature of the reaction in the eighteenth century.

We consider 1814 as the starting point of catalysis, when Kirchhoff recognized that acids enable the hydrolysis of starch to glucose. A few years later, Sir Humphry Davy discovered that a hot platinum wire lights up when brought into contact with a mixture of coal gas and air; there was oxidation but no flame and the platinum was unchanged. His cousin, Edmund Davy, found a few years later that finely divided platinum exhibited the same phenomenon at room temperature. This discovery led to the invention of Davy's mine safety lamp which, in probably the first practical application of heterogeneous catalysis, prevented the explosion of mine gas. Another early application of this discovery was a lighter, known as Döbereiner's Tinder Box, which is said to have sold over a million pieces!

In 1818 the French scientist Louis Thénard discovered hydrogen peroxide, and studied its decomposition over several materials. Thénard was also among the first to discover that metals such as iron, copper, and platinum induce the decomposition of ammonia (1913). Thénard's work on H_2O_2 decomposition is of particular interest because it represents one of the earliest measurements of the rate of a reaction.

Most catalytic reactions at that time were oxidations over platinum. Johann Döbereiner was the first to do a selective oxidation; in 1823 he found that ethanol reacts with oxygen over platinum to form acetic acid. Peregrine Philips patented

TABLE 1.2. Scientific Developments in Catalysis and Chemical Kinetics until 1940

Year	Name	Development
<1800		Several reports of reactions demonstrating, in hindsight, catalytic features
1813	Thénard	Ammonia decomposition over Fe, Cu, Ag, Au, and Pt
1814	Kirchhoff	Hydrolysis of starch to glucose catalyzed by acids
1817	H. Davy	Mixture of coal gas and air makes a platinum wire white hot
1818	Thénard	Measurements on the rate of H_2O_2 decomposition
1823	Döbereiner	Selective oxidation of ethanol to acetic acid over platinum
1834	Faraday	Comprehensive paper on the $H_2 + O_2$ reaction on platinum foils, including reaction rates, deactivation, reactivation, and poisoning
1835	Berzelius	Definition of catalysis, catalyst, and catalytic force
1850	Wilhelmy	First quantitative analysis of reaction rates
1865	Harcourt, Esson	Systematic studies on the concentration dependence of reaction rates
1884	van't Hoff	First concise monograph on chemical kinetics
1887	Ostwald	Definition of order of reaction
1889	Arrhenius	Arrhenius equation: $k = \upsilon \exp(-E_a/RT)$
1905	Sabatier	Hydrogenation of unsaturated hydrocarbons on nickel, start of organic heterogeneous catalysis
1905	Nernst	Third law of thermodynamics
~1909	Haber	Ammonia synthesis
1913	Chapman	First application of the steady-state approximation
1915	Langmuir	Quantitative theory of adsorption; description of catalysis in terms of a layer of gas molecules held on a surface, by the same forces that held atoms of a solid together
1921	Lindemann	Mechanism of unimolecular reactions
1925	Taylor	Catalytic reactions occur on active centers at surfaces; distinction between physisorption and chemisorption
1927	Hinshelwood	Kinetic mechanisms of heterogeneous catalytic reactions: Langmuir–Hinshelwood kinetics
1931	Onsager	Basis of non-equilibrium thermodynamics
1932	Eyring and Polanyi	Potential energy surface for the reaction $H + H_2$
1935	Eyring; Polanyi and Evans	Transition-state theory

the oxidation of SO_2 to SO_3 on platinum, which was put into practice 44 years later in a sulfuric acid plant operating according to the contact process.

Jöns Berzelius noticed the common factor in these studies. In his annual review of chemistry of 1835 he defined catalysis as a *chemical event that changes the composition of a mixture, but not the catalyst*. His explanation of catalysis, however, was largely metaphysical: "Catalytic force actually means that substances are able to awaken affinities which are asleep at this temperature by their mere presence and not by their own affinity."

In this respect, Michael Faraday's view on the formation of water from hydrogen and oxygen on platinum foils was more realistic, considering the level of knowledge in 1835. He thought that hydrogen and oxygen were condensed on the surface of the platinum and that the reaction resulted from the mere proximity of

TABLE 1.3. Developments in Industrial Catalysis

1740		Sulfuric acid production in the Bell process (later lead chamber process)
1823	Döbereiner	First technological application of heterogeneous catalysis in Döbereiner's Tinder Box
~1825	E. Davy	Mine safety lamp
1831	Philips	Patent for oxidation SO_2 to SO_3 over platinum (British Patent No. 6096)
1860s		Deacon process: oxidation of HCl to Cl_2 over copper chloride
1869	von Hoffmann	Partial oxidation of methanol to formaldehyde on silver
1875	Messel	Industrial oxidation SO_2 to SO_3 over platinum
1889	Mond	Discovery of steam reforming of hydrocarbons over nickel to CO and H_2
~1904	Ostwald and Brauer	Oxidation of ammonia to nitric oxide over platinum (Ostwald process)
1909	Haber, Bosch, Mittasch	Industrial development ammonia synthesis
1913	Bergius	First ammonia synthesis plant opened by BASF at Oppau, Germany.
		Catalytic coal liquefaction (applied in 1930s and 1940s)
1923	Fischer and Tropsch	Discovery of hydrocarbon synthesis from CO and H_2 on iron catalysts
		First methanol synthesis plant, opened by BASF at Merseburg, using a zinc chromite catalyst
1931	Lefort	Heterogeneous epoxidation of ethylene on silver catalyst
1935	Ipatieff	Alkylation of paraffins with superacid catalysts (commercialized for the production of kerosene in 1942)
1936	Houdry	Catalytic cracking of oil in USA
		Fischer–Tropsch plants in Germany produce synthetic fuels from coal-derived synthesis gas
1938	Reppe	Reactions of acetylene (from coal) with CO and water or alcohols to acrylic acid and esters catalyzed by transition metal carbonyls
	Roelen	Hydroformylation or oxosynthesis: reaction of CO and olefins to aldehydes catalyzed by cobalt carbonyls; the work of Reppe and Roelen marks the start of homogeneous catalysis
1946		Selective catalytic oxidation (e.g., o-xylene to phthalic anhydride)
1950		Catalytic reforming of naphta to gasoline over platinum-based catalysts
1955	Ziegler, Natta	Polymerization catalysis: high-density polymers from olefins over $TiCl_4/Al(C_2H_5)_3$ complexes at low pressure
1957	Smidt and coworkers	Wacker process: acetaldehyde from ethylene catalyzed by $PdCl_2/CuCl_2$ complexes in water
1955–1959		Synthetic zeolites X, Y, and A
1962	Weisz	Discovery of catalytic shape selectivity of zeolites
1963		Ammoxidation of propylene to acrylonitrile
1964	Rosinsky and Plank	Catalytic cracking (zeolites)
	Banks and Bailey	Olefin methathesis
		Oxychlorination
~1971	Paulik and Roth	Methanol carbonylation to acetic acid catalyzed by rhodium complexes in solution
1976		Catalytic automotive emission control
1977		Zeolites introduced in alkylation processes
1980s		Gasoline from methanol over H-ZSM-5 zeolites

the reactants. Not knowing that hydrogen and oxygen were diatomic molecules, Faraday could not surmise that platinum-induced dissociation is the key feature in the catalytic $H_2 + O_2$ reaction.

In the fifty years that followed, discoveries continued. Some catalytic reactions were applied industrially, such as the Deacon process (the oxidation of HCl to chlorine over copper chloride) and the oxidation of sulfur dioxide to sulfur trioxide.

At the end of the nineteenth century a rich chemical knowledge existed on catalysis. However, it was physical chemistry that enabled the large scale application of heterogeneous catalysis in practical processes, starting with the ammonia synthesis.

1.3. KINETICS AND CHEMICAL THERMODYNAMICS

According to Ostwald, a catalyst affects only the rate of a chemical reaction and not its equilibrium.

Kinetics—the measurement and analysis of rates of reaction—also finds its origin in the early nineteenth century. We already mentioned Thénard's measurements of the rate of hydrogen peroxide decomposition, around 1818. The first systematic analysis of reaction rates is attributed to Ludwig Wilhelmy. Around 1850 he measured the rate of inversion of cane sugar in the presence of different acids, by means of a polarimeter. He noted that the rate of change in sugar concentration was proportional to the concentrations of both the sugar and the acid. To analyze the data, he set up differential equations for the reaction rate. He even proposed an empirical relation to express the influence of temperature on the rate of reaction.

In 1862, the French chemists Pierre Berthelot and Léon de Saint-Gilles reported on the reaction between ethanol and acetic acid to ethyl acetate and water. They found that the rate of product formation was proportional to the product of the reactant concentrations. In the same years, the Norwegian chemist Waage and the mathematician Guldberg formulated their law of mass action, which, in hindsight, was based on invalid kinetic procedures, although the result was correct.

Noteworthy is the work of Augustus Harcourt and William Esson in England between 1865 and 1867. The chemist Harcourt made detailed experiments on the reactions between hydrogen peroxide and hydrogen iodide as well as between potassium permanganate and oxalic acid, paying particular attention to the influence of the reactant concentrations on the rate. The mathematician Esson analyzed his results in terms of differential equations which are similar to the ones used today. Esson was also the first to treat the kinetics of consecutive reactions, a topic we discuss in the next chapter.

In 1884, van't Hoff published a monograph on chemical kinetics, dealing with rate equations of reactions in the gas phase, in solutions, and at surfaces. In thermodynamics he derived equations for the effects of temperature and pressure on the equilibrium constant of a reaction. Arrhenius, in 1889, interpreted one of van't Hoff's relations in terms of an activation energy, which largely determines the rate of the reaction. The expression has become known as the *Arrhenius equation*. Friedrich Ostwald was the first to realize that a catalyst cannot change the equilibrium of a reaction, but that it accelerates the reaction by lowering the activation energy. These three scientists, van't Hoff, Ostwald, and Arrhenius, laid the foundations of physical chemistry. Interestingly, their work was based on a concept of matter which was far from the atomistic and molecular principles of statistical thermodynamics, which later provided a more fundamental basis to the Arrhenius equation in the 1930s with the formulation of the transition-state theory.

Ostwald clearly had the ambition to formulate general laws for chemical transformations, analogous to the fundamental laws of physics. Classical physics had become remarkably successful by the end of the nineteenth century. It was beginning to have a major impact on the daily life of citizens, especially due to the introduction of electricity and the rapid growth of transport. Thermodynamics was closely related to the development of the steam engine, and steam locomotives were becoming essential to travel. Classical physics had been expressed in the form of general laws, e.g., Newton's equation of motion, Ohm's relation between voltage and current, and, closer to chemistry, the laws of thermodynamics. Their generality was considered essential to their success and had raised physics in high esteem.

Chemistry was in a much more awkward stage. Although the industrial production of synthetic chemicals which replaced many natural products was significant, the underlying synthetic organic chemistry was predominantly an empirical science at that time. Ostwald considered it the task of physical chemistry to provide general laws for chemical transformations in a generally applicable, macroscopic formalism, analogous to what the now classical laws had accomplished in physics. The development of chemical thermodynamics by van't Hoff was an important outcome of such attempts. It provided a rigorous theory for predicting the temperature and pressure at which a chemical reaction occurs, as well as the maximum conversion levels that can be achieved, once the thermodynamic properties of reactants and products are known.

Around 1910, scientists became aware that many chemical reactions proceed through a series of consecutive elementary reaction steps. The tool for analyzing the kinetics of such mechanisms is the steady-state approximation, which was first applied by David Chapman in 1913. Bodenstein, however, became the real advocate of the steady-state treatment. He applied it widely in his work on gas-phase reactions and on reactions at surfaces. He ably defended the steady-state approximation against criticism. Of particular importance with regard to catalytic reactions is the work of Cyril Hinshelwood at Oxford in the 1920s. His studies on reactions at surfaces were largely inspired by Irving Langmuir's adsorption theory and contributed greatly to the understanding of catalytic reaction mechanisms. We shall discuss Langmuir–Hinshelwood kinetics in Chapter 2.

Arrhenius' rate expression and concept of an activation energy provided an important basis for the analysis of the rate of chemical reactions. However, the main difficulty that remained was the absence of a general theory to predict the parameters in the rate expression. Whereas equilibria of reactions could be rigorously defined, the determination of reaction rates remained a branch of science, for which the basic principles still had to be formulated. This was achieved in the 1930s, when Henry Eyring, and independently, Michael Polanyi and M. G. Evans, formulated (and later refined) the transition-state theory. An important aim of this book is to present the current understanding of the Arrhenius equation and its parameters in the context of catalytic reactions.

The Nobel prizes awarded in the field of physical chemistry, kinetics, and catalysis (Table 1.4) reflect the ongoing scientific progress in these fields. Nobel prizes for heterogeneous catalytic reactions have been awarded in the early part of this century. Many current major catalytic processes still owe their origin to that period. In the middle of this century, organometallic systems and inorganic coordination complexes formed the basis of catalytic innovations. In the last two decades, the application of zeolites has been the basis of several new processes.

When van't Hoff received his Nobel prize in 1901, the study of chemical equilibrium thermodynamics was almost complete. Kinetics, however, belongs to the field of non-equilibrium thermodynamics, a subject for which the principles still had to be formulated. The 1931 work of Lars Onsager marks the beginning of the linear non-equilibrium thermodynamics. This discipline provides a firm basis for the kinetics of the steady state, which applies to many catalytic processes. Onsager received the Nobel prize in 1968. Recently, oscillating reactions have

TABLE 1.4. Nobel Prizes in Chemistry, Awarded for Work in the Fields of Catalysis, Chemical Kinetics, and Physical Chemistry

1901	J.H. van't Hoff	Physical chemistry
1903	S.A. Arrhenius	Physical chemistry
1909	F.W. Ostwald	Catalysis, reaction rates
1912	P. Sabatier	Catalytic hydrogenation
1919	F. Haber	Ammonia synthesis
1920	W.H. Nernst	Physical chemistry
1931	F.C.R. Bergius, C. Bosch	Chemical engineering
1932	I. Langmuir	Surface chemistry
1956	C.N. Hinshelwood, N.N. Semenov	Chemical kinetics
1963	K.W. Ziegler, G. Natta	Polymerization catalysis
1967	M. Eigen, R.G.W. Norrish, G. Porter	Kinetics, flash photolysis
1968	L. Onsager	Irreversible thermodynamics (linear)
1973	E. Fischer, G. Wilkinson	Organometallic chemistry
1977	I. Prigogine	Irreversible thermodynamics (non-linear)
1986	D.R. Herschbach, Y.T. Lee, J.C. Polanyi	Reaction dynamics
1992	R.A. Marcus	Electron transfer reactions

attracted the attention of the catalytic community. Such reactions proceed under conditions that are far from equilibrium and belong to the field of non-linear thermodynamics. The theory behind the "far-from-equilibrium thermodynamics" has largely been developed by Ilya Prigogine, who received the 1977 Nobel prize in chemistry for his work.

Recent major advances in kinetics relate to the development of experimental and theoretical methods that probe molecular events on time scales comparable to the life times of reaction intermediates. Molecular beam techniques, as applied in modern reaction dynamics, furnish detailed information on the potential energy surfaces of the interacting reactants, which can be used to compute in detail the time-dependent behavior of the reactants.

1.4. THE AMMONIA SYNTHESIS

Chemical equilibrium thermodynamics provided the basis for the development of new catalytic processes in the beginning of this century.

At the turn of this century, people became aware that the increase in population growth, when compared to the limited area of land available for wheat growing and a relatively low yield per acre of soil, would inevitably lead to famine in the future, unless fertilizers, such as ammonium sulfate, were used on a large scale. Also, politicians argued that the production of explosives should not depend on the vulnerable import of Chilean guano (sodium nitrate). It was obvious that nitrogen from the air represented a virtually unlimited resource, and fixation of atmospheric nitrogen became a major challenge for chemistry. It was the main driving force for the development of the ammonia synthesis.

Although the catalytic decomposition of ammonia had been known since the early nineteenth century, attempts to synthesize ammonia from its elements had failed. This of course is not surprising, as these experiments were done at atmospheric pressure, where the equilibrium concentration of ammonia at around 500°C is only about 0.1%. The newly acquired insights in equilibrium thermodynamics enabled Fritz Haber and Walther Nernst to predict that the synthesis of ammonia would only be possible at high pressures and moderate temperatures. Whereas Nernst openly expressed serious doubts about the feasibility of such a process, Haber believed that it was possible. However, this necessitated the invention of a highly active catalyst, as well as the development of new engineering technology for carrying out reactions in continuous flow operation at high pressures.

In 1908, Haber, a professor at the University of Karlsruhe, Germany, started a collaboration with the chemist Carl Bosch of the Badische Anilin- und Soda-Fabrik (BASF) at Ludwigshafen. Their 1909 experiment in which 8% of ammonia formed at 550°C and 175 atm confirmed the predictions from chemical thermodynamics and convinced them of the feasibility of ammonia synthesis at an industrial scale.

The same year Alwin Mittasch began systematic tests of an incredible horde of catalysts in a reactor operating at 200 atm. At the beginning of 1912, 6500 tests already had been performed based on 2500 different catalyst formulations. The choice was made for an alkali-promoted, fused-iron catalyst which contained several other oxidic promoters.

On September 9, 1913 the first ammonia synthesis plant was opened at Oppau, Germany, near Ludwigshafen, with a production of 10,000 kg of NH_3 per day. The plant was designed for the production of ammonium sulfate, a fertilizer.

During the first world war, however, Germany was entirely cut off from the import of guano, used as raw material for the fabrication of explosives. Soon the German government realized that nitrates could very well be made by the Ostwald process. This was done by oxidizing ammonia over platinum to NO, which is subsequently oxidized to NO_2 and then reacted with warm water to form nitric acid. Consequently, the government summoned the further development of nitrate production via the Ostwald route. From early 1915, a new reactor operating with a cheap iron–manganese–bismuth oxide catalyst started to produce nitric acid, with ammonia from the Haber–Bosch process. In later years the capacity was greatly enlarged. If Germany had not had the ability to use ammonia synthesis and the Ostwald process to produce nitrate and explosives, World War I might have ended earlier than 1918.

The realization of the ammonia synthesis in a practical process represents an impressive accomplishment, particularly if one considers that many other problems had to be solved in addition to finding an active catalyst. Essential for a large scale process, it can be carried out in a continuous operation, with a solid catalyst in a flow reactor. Unlike in homogeneous batch processes, a step to separate the catalyst from reactants and products is not required. Also, the conditions needed for ammonia synthesis demanded the development of high-pressure reactors and special precautions with respect to cooling. Gas separation technology had to be employed for the production of nitrogen, while hydrogen was produced in additional catalytic processes. In short, many new engineering practices had to be developed. In this way heterogeneous catalysis started chemical engineering as we know it today.

1.5. CATALYSIS AND THE GROWTH OF THE CHEMICAL INDUSTRY

Both catalysis and process engineering have formed the basis of innovations in the chemical industry.

The tremendous progress in reaction technology and in heterogeneous catalysis caused by the ammonia synthesis opened the way for many other new processes. These were mainly based on hydrogenation reactions. Base chemicals could now be produced at a large scale. For instance, the catalytic synthesis of methanol from

CO and H_2 requires conditions that are rather similar to those of the ammonia synthesis. The production of synthesis gas, CO + H_2, by steam reforming of hydrocarbons had been known since 1888, and was already used as the source of hydrogen in the ammonia synthesis. Methanol production commenced in 1923.

Hydrogenation of unsaturated hydrocarbons as well as of carbon monoxide over nickel catalysts had been discovered by Paul Sabatier in 1905, which may represent the start of organic heterogeneous catalysis. It was rapidly applied in the food industry for the hydrogenation of vegetable oils to make margarine.

In 1923, Franz Fischer and Hans Tropsch reported the synthesis of long chain hydrocarbons from synthesis gas over alkali-promoted iron catalysts. The Fischer–Tropsch process became operational in 1936, and made Germany independent of imported fuels in World War II.

Direct hydrogenation of coal into liquid hydrocarbons was developed in 1936 by Friedrich Bergius. Catalytic hydrodesulfurization and hydrogenation processes using sulfidic catalysts were used to remove sulfur from the oil generated by hydrogenative coal liquefaction. The chemical industry relied heavily on coal and on coal-derived acetylene. Coke heated with calcium oxide forms calcium carbide, which reacts with water to form acetylene and calcium hydroxide. Acetylene has a rich chemistry, and served as a basic chemical building unit for the organic chemical industry. Between 1938 and 1945, Walter Reppe discovered a wealth of reactions between unsaturated hydrocarbons such as acetylene, ethylene, and higher alkenes with carbon monoxide in water or alcohol using transition metal carbonyls as catalysts. A typical example is the reaction of acetylene with carbon monoxide and water over nickel tetra carbonyl as the catalyst forming acrylic acid ($CH_2{=}CHCOOH$). The chemistry developed by Reppe, together with the hydroformylation to be discussed later, marks the beginning of an extremely important subfield: homogeneous catalysis.

The coal-based chemical industry was gradually replaced by a petrochemical one beginning around 1935. Catalytic cracking of petroleum over acid-treated montmorillonite-type clay catalysts was implemented by Eugene J. Houdry (a mechanical engineer and an automobile race driver) in the United States in 1936. The high demands for aviation fuel in the second world war led to a rapid expansion of the cracking process. The clays were gradually replaced by amorphous silica–aluminas, and finally by synthetic zeolites in the sixties. Catalytic cracking grew into one of the largest catalytic processes.

The shift from coal to oil, of course, had major consequences for the chemical industry. Alkenes, instead of synthesis gas and acetylene, became the basic building blocks. Homogeneous catalysis became increasingly important for the production of chemicals. We already mentioned the Reppe chemistry, based on acetylene as the main feedstock. In 1938, the same year that Reppe started his work, another German scientist, Otto Roelen, discovered the reaction of olefins with carbon monoxide and hydrogen over a cobalt carbonyl catalyst to form aldehydes, known as oxo synthesis or, known more commonly as hydroformylation. Among the major achievements that followed were the development of the Wacker process, the oxidation of ethylene to acetaldehyde on palladium chlorides (Jürgen Smidt and coworkers, 1957–1959), the carbonylation of methanol to acetic acid with rhodium complexes by Frank E. Paulik and James F. Roth in the late 1960s, and, very importantly, all kinds of homogeneous polymerization reactions.

The first processes for polymers and synthetic rubber were developed in the 1940s, prompted by the second world war when a replacement for natural rubber was needed. The early olefin polymerization technology involved radical reactions at pressures around 2000 bar, which essentially represented the inverse of thermal cracking. The breakthrough in this field came with Ziegler–Natta catalysis in the mid-1950s. In this process the polymerization of alkenes is catalyzed by $TiCl_4/Al(C_2H_5)_3$ complexes in solution, under mild conditions (low pressure and temperature). The reaction yields polymer products of significantly higher density, and correspondingly, higher melting point and crystallinity than do the free radical processes.

Since the 1950s, many processes based on homogeneous reactions have been introduced, either to replace heterogeneous catalytic reactions with routes offering milder and less expensive process conditions, or to make entirely new products. Fine chemicals in the pharmaceutical industry form an outstanding example of the latter. Nevertheless, heterogeneous catalysis also expanded immensely.

After World War II, the growth of automobile and aircraft transportation stimulated research on petroleum conversion processes. New synthetic fuel-producing processes were introduced that enabled the interconversion of oil fractions obtained by destillation. These processes often used catalysts based on noble metals as well as solid acids, like the zeolites, applicable at high temperatures. Superacids were used as catalysts for alkylation to branched olefins for high-octane kerosene. Catalytic reforming with noble metals on acidic supports was introduced in the production of gasoline. Catalytic cracking became a major application of solid acids.

The 1960s brought significant improvements in oil refining, with the introduction of a new generation of cracking catalysts based on zeolites. These materials have well-defined pore structures, and provide size restrictions to the products, in addition to having the catalytic properties of solid acids. Their narrow micropores provided the main reason for their improved stability, owing to the suppression of coke formation. Another advancement was the use of bimetallic platinum–rhenium and platinum–iridium catalysts, which offer considerably improved stability against coke formation in gasoline reforming.

New selective catalytic oxidation processes, such as the oxidation of propylene to acrolein, or the ammoxidation of propylene to acrylonitrile catalyzed by mixed oxides, were developed to produce monomers for a new generation of polymers in the 1960s.

In the 1970s, the oil crisis generated a need for alternative raw materials. Coal and especially natural gas were reconsidered as carbon sources. Synthesis gas, now produced from natural gas, became an alternative feedstock for the production of oxygenated hydrocarbons. For instance, a new process employing organorhodium compounds was developed to produce acetic anhydride from synthesis gas via carbonylation of methyl acetate.

Environmental concerns led to the development of gas treatment catalysts, not only to reduce emission from stationary sources such as the stackgas of electricity generation plants, but also to reduce the emission of NO and CO in the exhaust gases from automobiles. New heterogeneous catalysts were applied: supported oxides of vanadium and molybdenum for removing nitrogen oxides from the stack gas of power plants, and monolith-supported noble metals such as platinum, rhodium and palladium in the three-way catalyst to reduce the emission of NO, CO, and hydrocarbons from cars. The latter application represents a remarkable highlight in the history of catalysis, as it has grown in less than two decades to the largest segment in the catalyst manufacturing market.

Whereas such applications of catalytic "clean-up" technology are of obvious importance, it is even more desirable to develop new catalytic processes that produce less or no waste products. Many conventional routes consume acids or bases and produce salts which cause waste problems. The production of ethylene epoxide from ethylene by direct oxidation over silver catalysts forms an example for a clean, catalytic process which has replaced traditional routes involving the

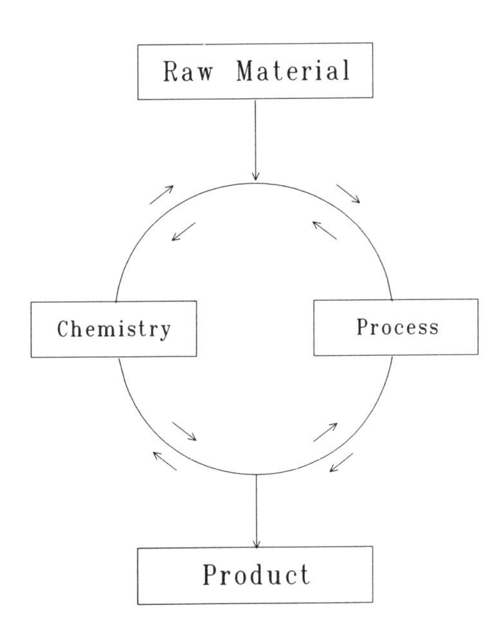

Figure 1.1. The innovation loop: The success of a process depends on the performance of a catalyst. Operational constraints require the development of an improved catalyst, resulting in a higher process efficiency.

consumption of Cl_2 and the production of large amounts of $CaCl_2$. New catalytic applications are also being developed to make products that are less harmful to the environment. Replacement of ozone-degrading chlorinated refrigerants by less active fluorided hydrocarbons is a recent example. New catalysts based on solid Lewis acids have been developed for this purpose.

Developments in catalysis and chemical process engineering have been linked in an innovation loop as sketched in Figure 1.1. The cycle starts with society, which generates a need for new products or requests a change in the use of raw materials. The implementation of a new or improved process and the performance of a particular catalyst are mutually dependent. Catalyst improvements may result in a significantly increased process efficiency, while process constraints dictate the improvements required of a particular catalyst. The cycle is closed with new or more economically manufactured products, available to society. Table 1.1 gives a historical perspective of the changes that occurred in the chemical industry while Table 1.3 lists the highlights in industrial catalysis.

1.6. THE SCIENTIFIC DISCIPLINES OF CATALYSIS

Kinetics, synthesis, and characterization form the three subdisciplines of catalysis. Physical techniques offer means to relate catalyst performance with structure and composition. Theory and experiment provides an in-depth understanding of catalysis on a molecular level.

Catalysis as a science has evolved along three major directions: kinetics, catalyst characterization, and the synthesis of catalysts. These three ingredients can all be recognized on three different levels, as schematically summarized in Figure 1.2. Catalysis at the microscopic level deals with elementary reaction steps of molecules on the surface of the catalyst, spectroscopic and theoretical studies on surface species and reactive sites, and the synthesis of active sites and reaction intermediates. The mesoscopic level is that of measuring catalytic activities and selectivities, characterization of catalysts by spectroscopy, diffraction, microscopy and temperature–programmed methods, and of catalyst preparation by impregnation or precipitation, followed by steps such as calcination, reduction, or sulfidation. The macroscopic level is that of reaction engineering, texture determination and shaping catalysts in extrudates, as powders for fluid beds, or in monoliths. In this book, we are interested in the microscopic level.

Kinetics

In practical catalysis, the chemical engineer considers catalytic performance defined once the kinetics have been established. It is relevant to note at this point the difference between intrinsic and extrinsic kinetics of a catalytic process. The

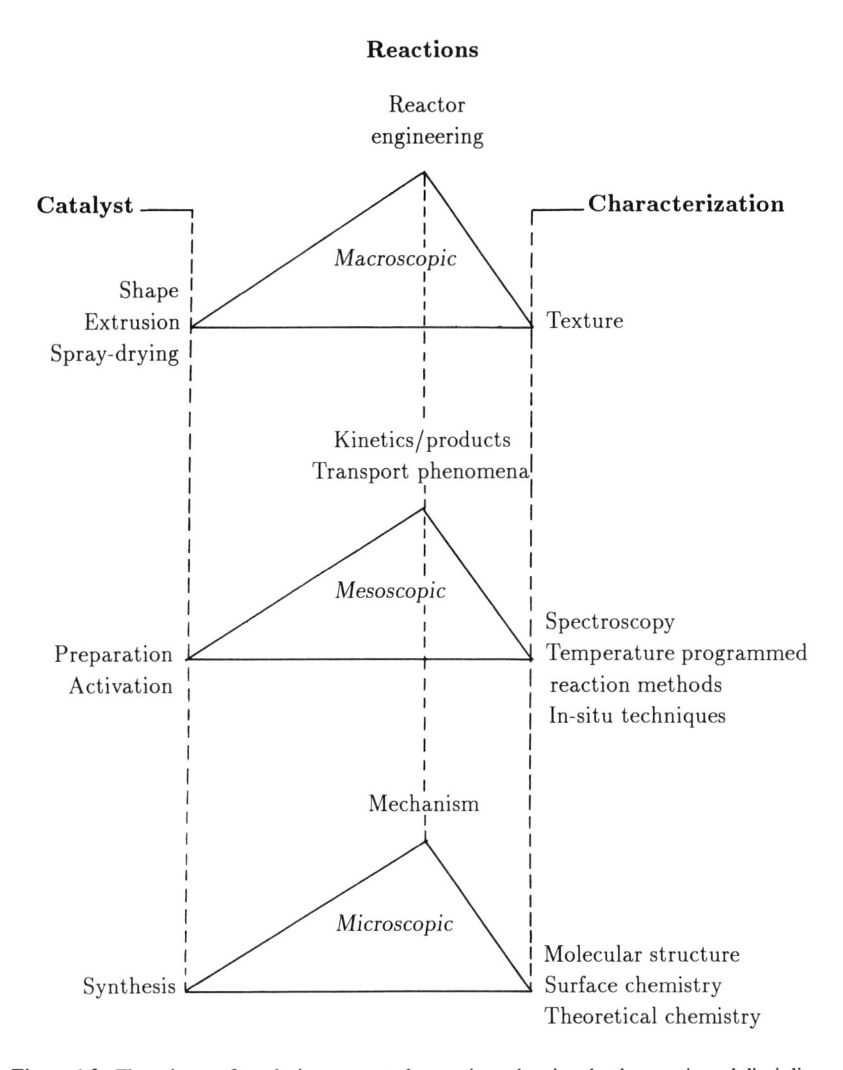

Figure 1.2. The science of catalysis represented as a prism, showing the three major subdisciplines of catalysis on the macro-, meso-, and microscopic level (from van Santen, 1991).

overall rate of a process is controlled by the rate at which mass and heat are introduced and transferred from the reacting system, and by the rate of the chemical transformation of the reactants by contact with the catalyst. Clearly, the reactor itself as well as the shape and porosity of the catalyst particles may have a large effect on mass and heat transport properties. These physical properties control the so-called extrinsic kinetics of a catalyst; these are obviously very important to the catalytic chemical engineer. Here we focus on the intrinsic or chemical kinetics, defined by the rates of the chemical transformations themselves.

Chemical kinetics deals with the rate at which chemical transformations take place. As discussed in Chapter 2, the conversion of reactants A and B to a product P occurs usually in a number of consecutive elementary reaction steps. The mechanism leads to a rate equation, which may, for instance, be of the form

$$r = \frac{d[P]}{dt} = k\,[A]^x\,[B]^y \tag{1.1}$$

in which k is the reaction rate constant, $[A]$, $[B]$, and $[P]$ are the concentrations of reactant and product molecules, and x and y the orders of the reaction. Rate equations are discussed in detail in Chapter 2. One of the main tasks in catalysis is to relate rate parameters k, x, and y to physicochemical properties of the catalyst.

Initially, kinetics became a major tool for deducing mechanistic information. The relation between the mechanism of a reaction and kinetics forms an important subject of this book. Rate equations such as (1.1) follow from a postulated mechanism. However, the fact that a mechanism may be able to explain kinetic experiments by no means implies that this particular mechanism is necessarily correct. It appears that many different mechanisms may result in similar kinetic behavior. Thus, additional information on the reacting species on the surface of the catalyst is necessary to reach a definitive understanding of catalysis in relation to the structure and composition of the catalyst.

Characterization

Heterogeneous catalysts are chemically complex materials. As catalysis occurs at the surface, one preferentially uses catalysts in which the active component is present in the form of small particles. Such particles are rarely stable, tend to agglomerate, and moreover, densely packed beds of small particles generate high pressure gradients in the reactant flow through a reactor, so practical catalysts usually consist of an inert, porous support with a high surface area with the catalytically active components dispersed on it as small particles. Catalysts operating in industrial processes often have an even more complex composition due to the presence of promoters, or to contaminants that are introduced with the reactants.

Most of the early progress in fundamental catalysis can be attributed to new characterization tools, introduced with the main objective to establish correlations between catalyst performance and catalyst structure. The techniques developed in the first part of this century exploited physical or chemical adsorption and provided information on the surface area of the catalysts. Only recently, techniques have become available that enable analysis of the structure of surfaces at an atomic level. The application of modern spectroscopic techniques has made a tremendous impact, not only on catalyst characterization but also on the study of reaction mechanisms, as intermediates formed at the catalyst surface can be studied under in situ conditions.

Progress in fundamental science depends strongly on the design of well-defined model systems. New ultrahigh vacuum technology that became available on a large scale in the 1960s and 1970s opened the way for the development of surface science. The well-defined surfaces of single crystals have become highly appreciated models of catalysts, which provided detailed information on the structure and bonding of adsorbates on metals. Experiments with metal clusters distributed in a molecular beam have given insight into the reactivity of metal clusters as a function of their size. These developments have provided the investigator in modern catalysis with the tools to generate a depth of understanding of catalysis on a molecular level, as has been traditionally more common in other fields, such as organic chemistry.

In order to deduce structural information from spectroscopic data, one needs theoretical methods. Advanced computational methods of theoretical chemistry and powerful computer hardware are being applied increasingly in catalysis in predictive studies of chemical reactivity.

Catalyst Synthesis

Also in catalyst synthesis there is a change in techniques that enable control of the reaction at the molecular level. The main aim of catalyst preparation techniques applied in heterogeneous catalysis is to disperse the catalytically active component over a large surface area of a supporting material. This discipline can be considered a part of applied colloid and inorganic chemistry. In addition, the shape and porosity of catalysts need to be controlled in order to satisfy requirements in reaction engineering. The importance of zeolite catalysts has generated new techniques for catalyst synthesis based on sol-gel chemistry. Traditionally, the preparation of solid catalysts has been more art than science. However, the application of modern characterization methods and model systems is providing this field with a scientific basis.

Coordination chemistry and organometallic chemistry enable molecular control of catalytic complexes by manipulation and choice of ligands. Such complexes can also be used as precursors to the catalytically active components of heterogeneous catalysts, resulting in homogeneous distribution of well-defined catalytic clusters. These developments create the promise of a molecular catalytic engineering approach.

1.7. THE SCOPE OF THIS BOOK

Rates of surface reactions can be predicted based on an understanding of the reactivity of intermediates formed on the catalyst surface. Transition-state theory enables the prediction of the constants in Arrhenius's rate expression.

The availability of powerful computers and advanced computational methods to treat problems in chemistry opens the possibility for predicting rates of reactions. As explained earlier, equilibrium thermodynamics has provided a rigorous basis for the prediction of maximum conversion levels and the conditions under which they are achieved. The Arrhenius equation served as a tool for rationalizing rate constants in terms of activation energies and preexponentials. These parameters, however, could not be predicted on the basis of molecular properties of the reacting species until the concept of the transition state evolved, around 1935. Gas-phase kinetics in particular established a fundamental understanding of the Arrhenius parameters. We treat the transition-state theory in Chapter 4.

A reaction between molecules implies the breaking of existing bonds and the formation of new bonds. The energy of the participating molecules changes continuously during the reaction event, the latter being commonly represented as a pathway over a potential energy surface. The reacting molecules follow that reaction path which leads over the lowest energy barrier separating reactants and products. It appears that the structure of the reaction intermediate at this barrier, together with details of the energy change around the barrier, controls the rate of a reaction to a significant extent. The transition-state theory furnishes the equations for deducing the rate of reaction, provided information on the transition state is available. The need to make assumptions about the transition state has, for a long time, been the major weakness of the theory. At present, one can use computational methods to construct potential energy surfaces, which are based on increasingly accurate calculations of the electronic structure in interacting molecules. This, in principle, generates the possibility to predict reaction-rate constants based on atomistic information of surface complexes that are involved in the reaction.

In this book, we demonstrate the use of transition-state theory to describe catalytic reactions on surfaces. In order to do this we start by treating the kinetics of catalytic reactions (Chapter 2) and provide some background information on important catalytic processes (Chapter 3). In Chapter 4 we introduce the statistical mechanical basis of transition-state theory and apply it to elementary surface reactions. Chapter 5 deals with the physical justification of the transition-state theory. We also discuss the consequences of media effects and of lateral interactions between adsorbates on surfaces for the kinetics. In the final chapter we present the principles of catalytic kinetics, based on the application of material given in earlier chapters.

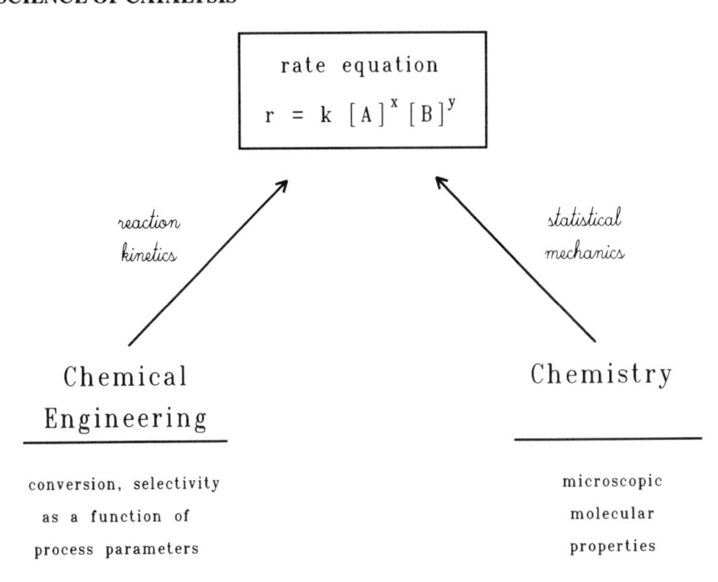

Figure 1.3. Kinetic analysis results in an expression for the rate of a reaction, while statistical mechanics provides the tools to predict the constants in the rate equation based on microscopic properties of the reacting molecules.

Figure 1.3 illustrates the concept on which this book is based. It shows the relation between macroscopic kinetics, as used by the chemical engineer, and microscopic atomic information, as needed or provided by the chemist. The connection is provided by the rate equation (1.1). Rates of catalytic reactions can be predicted from the reactivity of intermediates absorbed on the surface of the catalyst, using transition-state theory to calculate the parameters in the Arrhenius expression for the reaction-rate constant. This is the philosophy behind this book.

2

THE RATE EQUATION

2.1. THE REACTION EQUATION

If a system of reactants and products is not in equilibrium, then the reaction is driven by its chemical affinity. If the system is close to equilibrium, the reaction rate decreases exponentially to zero until equilibrium is reached.

Let us consider the conversion of molecules A and B into molecules C and D. The rate of disappearance of compound A follows from the chemical reaction equation of the system

$$aA + bB \underset{k^f}{\overset{k^b}{\rightleftharpoons}} cC + dD \tag{2.1}$$

in which k^f and k^b are rate constants, independent of concentration, of the forward and backward reactions. The numbers a, b, c, and d are stoichiometric coefficients.

A reaction which proceeds exactly as expressed by a stoichiometric reaction equation is called an *elementary reaction*. Thus, in the case of (2.1), when a collision between a molecules of A and b molecules of B leads to the production of c molecules of C and d molecules of D. The rate is given by

$$\begin{aligned} r = -\frac{1}{a}\frac{d[A]}{dt} &= k^f[A]^a[B]^b - k^b[C]^c[D]^d \\ &= r^f - r^b \end{aligned} \tag{2.2}$$

where $[A]$ denotes the concentration of molecule A, etc. The overall orders of the forward and the backward reactions r^f and r^b are $a + b$ and $c + d$, respectively. The order of the reaction with respect to a particular compound is defined as the power coefficient in the concentration in the power rate laws describing r^f and r^b in (2.2). Hence, the order of the forward reaction (2.1) of the elementary reaction in A equals a.

However, many reaction equations represent the overall result of a number of elementary steps. In those cases the reaction orders differ from the stoichiometric coefficients. As we will see later, orders then can be nonintegral or negative numbers.

When the reaction rate is zero, the reaction is at equilibrium. The mass action law of Guldberg and Waage is then valid:

$$K^{\text{eq}} = \frac{[C]_{\text{eq}}^c [D]_{\text{eq}}^d}{[A]_{\text{eq}}^a [B]_{\text{eq}}^b} = \frac{k^f}{k^b} \tag{2.3}$$

This expression relates dynamic quantities, the rate constants of the forward and backward reactions, with the equilibrium constant K^{eq}, and provides a link between kinetics and thermodynamics. The equilibrium constant can be computed from the change in the standard Gibbs free energy, ΔG^0, or the standard chemical potential μ_i^0 of all components i in the reaction

$$-\ln K^{\text{eq}} = \frac{\Delta G^0}{RT} = \frac{1}{RT}\{c\mu_C^0 + d\mu_D^0 - a\mu_A^0 - b\mu_B^0\} \tag{2.4}$$

where R is the gas constant and T the absolute temperature.

An important consequence of (2.3) is if for a system the rate constant is known for the forward reaction, the rate constant for the backward reaction can be calculated. For example, the rate of the reaction between nitrogen and hydrogen to ammonia is difficult to determine, but can easily be derived from the rate of the ammonia decomposition and the equilibrium constant for this reaction.

Let us return to the rate of reaction defined in (2.2). Using the expression for the equilibrium constant, the reaction rate can be written in a form such that a driving force for the reaction can be recognized. This driving force is called the *chemical affinity.* First, we write the reaction rate in a form which visualizes how far the system is from equilibrium

$$\begin{aligned} r = r^f - r^b &= r^f\left(1 - \frac{r^b}{r^f}\right) \\ &= r^f\left(1 - \frac{1}{K^{\text{eq}}}\frac{[C]^c[D]^d}{[A]^a[B]^b}\right) \\ &= r^f\left(1 - \frac{[C]^c}{[C]_{\text{eq}}^c}\frac{[D]^d}{[D]_{\text{eq}}^d}\frac{[A]_{\text{eq}}^a}{[A]^a}\frac{[B]_{\text{eq}}^b}{[B]^b}\right) \end{aligned} \tag{2.5}$$

It is useful to introduce stoichiometric coefficients v_i. They are defined by the reaction equation

$$\sum_i v_i X_i = 0 \tag{2.6}$$

where X_i stands for the change in reactant and product concentration. In this equation, v_i is positive for products and negative for reactants. If we write c_i for the concentration of compound i, the equilibrium constant becomes

$$K^{\mathrm{eq}} = \prod_i \left(c_i^{\mathrm{eq}}\right)^{v_i} \tag{2.7}$$

and the change in the standard Gibbs free energy

$$\Delta G^0 = \sum_i v_i \mu_i^0 \tag{2.8}$$

With these definitions, expression (2.5) becomes

$$r = r^f \left(1 - \prod_i \frac{(c_i)^{v_i}}{(c_i^{\mathrm{eq}})^{v_i}}\right) \tag{2.9}$$

Second, we are going to relate the rate of the reaction with the chemical potentials, μ_i. For a reaction at equilibrium

$$0 = \sum_i v_i \mu_i^{\mathrm{eq}} \tag{2.10}$$

We now assume that for a system on its way to equilibrium we may write

$$\mu_i(t) = \mu_i^0 + RT \ln c_i(t) \tag{2.11}$$

where both μ_i and c_i are considered to be a function of time; μ_i^0 is the standard chemical potential. According to nonequilibrium or irreversible thermodynamics, (2.11) is allowed for a system with a large number of molecules which changes on a time scale that is long with respect to the time scale of the collisions between the molecules. Then temperature and pressure remain locally well-defined properties, and the system can be considered to consist of units which are all internally in equilibrium.†

---

†Strictly speaking, the chemical potential is only defined as a time-independent quantity for a system at equilibrium. Chemical kinetics belongs to the discipline of irreversible, nonequilibrium thermodynamics. Whereas thermodynamics of reversible processes has a solid foundation, this is not the case for the thermodynamics of irreversible processes. At present, the latter is one of the frontier areas of chemical research. We will highlight these developments where relevant to chemical kinetics and catalysis.

In the nonequilibrium situation one can define a quantity called reaction affinity, A (not to be confused with the compound A we used before), which indicates how far the system is removed from equilibrium

$$\sum_i \nu_i \mu_i = -A \tag{2.12}$$

The relation between reaction affinity and concentrations becomes clear if one combines (2.10–2.12) into

$$\begin{aligned} -A(t) &= \sum_i \nu_i \mu_i(t) - \sum_i \nu_i \mu_i^{\text{eq}} \\ &= \sum_i \nu_i RT \ln \frac{c_i(t)}{c_i^{\text{eq}}} \end{aligned} \tag{2.13}$$

If we rewrite (2.13) as

$$-\frac{A(t)}{RT} = \ln \prod_i \left(\frac{c_i(t)}{c_i^{\text{eq}}} \right)^{\nu_i} \tag{2.14}$$

one can easily see that reaction affinity and reaction rate are related as

$$r = r^f (1 - e^{-A/RT}) \tag{2.15}$$

When the deviation from equilibrium is small, expression (2.15) becomes[†]

$$r = \frac{r^f A(t)}{RT} \tag{2.16}$$

and a linear relation between rate and chemical affinity is found. For this reason chemical affinity can be considered the driving force of the reaction.

We will now show that when the deviations of the concentrations from equilibrium are small, the reaction rate decreases exponentially until equilibrium is reached. As we will discuss later, the system then has the maximum possible entropy within its constraints and the free energy is at minimum.

The solutions of the rate equation close to equilibrium can readily be found in the following way. Define ξ as the extent of the reaction and δ as the deviation from equilibrium:

[†]We use the approximation $e^x \approx 1 + x$ for $x << 1$.

$$\xi = \frac{1}{\nu_i}(c_i - c_i^i); \quad \delta = \frac{1}{\nu_i}(c_i^{\text{eq}} - c_i) \tag{2.17}$$

in which c_i^{eq} is the equilibrium concentration, and c_i^i is the initial concentration. When ξ equals zero, all concentrations are equal to the initial values, c_i^i; when δ equals zero, the reaction is at equilibrium. The maximum values of these parameters are

$$\xi^{\max} = \delta^{\max} = \frac{1}{\nu_i}(c_i^{\text{eq}} - c_i^i) \tag{2.18}$$

and the reaction rate becomes

$$r = \frac{d\xi}{dt} = -\frac{d\delta}{dt} \tag{2.19}$$

If the reaction is not too far from equilibrium such that δ is small, the rate of reaction r can be expressed in δ by substituting terms as $[A]_{\text{eq}} - a\delta$ for the concentrations in (2.2). Using the binomial relation

$$(x+y)^N = \sum_{n=0}^{N} \frac{N!}{(N-n)!n!} x^n y^{N-n} \tag{2.20}$$

and ignoring all terms that are quadratic or higher in δ, the solution of (2.2) becomes

$$r = -\frac{d\delta}{dt} = \tau^{-1}\delta$$

$$\delta = \delta^{\max} e^{-t/\tau} \tag{2.21}$$

in which τ is a function of the equilibrium concentrations of A, B, C, and D, and of the reaction-rate constants k^f and k^b

$$\tau^{-1} = k^f (b^2[A]_{\text{eq}}^a[B]_{\text{eq}}^{b-1} + a^2[B]_{\text{eq}}^b[A]_{\text{eq}}^{a-1}) -$$

$$k^b (d^2[C]_{\text{eq}}^c[D]_{\text{eq}}^{d-1} + c^2[D]_{\text{eq}}^d[C]_{\text{eq}}^{c-1}) \tag{2.22}$$

Similarly, one finds for the extent of reaction ξ

$$\xi(t) = \xi^{\max}(1 - e^{-t/\tau}) \tag{2.23}$$

Close to equilibrium, the time dependence of the chemical affinity follows from the expression for A in (2.13), the definition of δ in (2.17) and its time dependence in (2.21)†:

---

†Use the approximation $\ln(1+x) \approx x$ for $x << 1$.

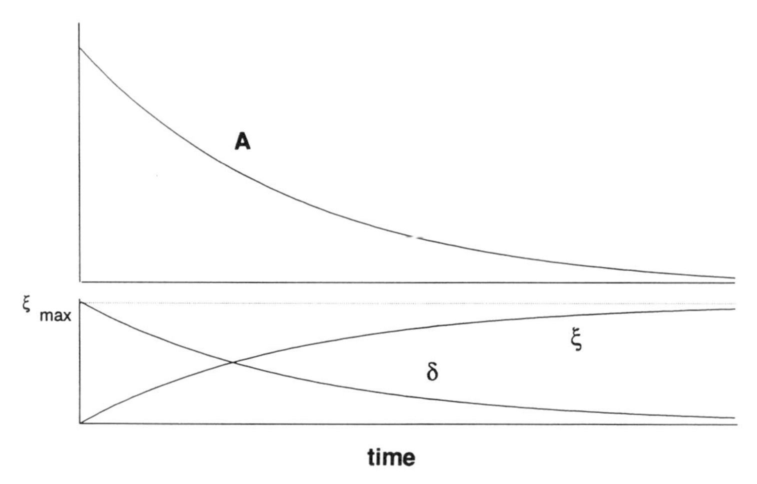

Figure 2.1. If the reaction is close to equilibrium, the extent of reaction, the deviation from equilibrium, and the chemical affinity all approach their equilibrium value in an exponential fashion.

$$A(t) = \delta^{\max} \sum_i \frac{v_i^2 RT}{c_i^{\text{eq}}} e^{-t/\tau} \tag{2.24}$$

The relaxation time of this exponential decay equals τ and is, according to (2.22), determined by rate constants, reaction orders, and equilibrium concentrations, but it does not depend on initial concentrations.

The important point to note is that every reaction which is close to equilibrium approaches the equilibrium state exponentially, irrespective of the overall reaction order (see Figure 2.1). Thus the reaction rate decreases exponentially to zero and the free energy decreases exponentially to its minimum, provided that the deviation from equilibrium is small.

2.2. THE SECOND LAW OF THERMODYNAMICS

A reaction system that is close to equilibrium has a positive entropy production.

In this section we will demonstrate that reaction-rate expressions such as (2.5) are consistent with the laws of thermodynamics. Thermodynamics enables the computation of differences in the state variables of a reaction system, before and after reaction. According to the first law of thermodynamics, *the total energy of a system is conserved*:

$$dE = dQ + dW - P\,dV \tag{2.25}$$

At constant volume V, the sum of heat content Q and work W is constant.

According to the second law of thermodynamics, *transformations occur such that the entropy of the system increases*:

$$dS = \frac{dQ}{T} \geq 0 \tag{2.26}$$

In order to derive the rate of the entropy change during a chemical reaction, we need a relation between local entropy and concentration. Again we require the assumption that the rate of change of the state properties is small compared to the rate of energy and mass exchange in volume units containing a number of atoms of the order of Avogadro's number. This is the local equilibrium assumption: Temperature and pressure remain locally determined variables.

The entropy can be related to the concentration by using the following form of the first law of thermodynamics:

$$dE = T\,dS - P\,dV + \sum_i \mu_i\,dn_i \tag{2.27}$$

The change in entropy follows directly from (2.27)

$$dS = \frac{dE}{T} + \frac{P}{T}dV - \frac{1}{T}\sum_i \mu_i\,dn_i \tag{2.28}$$

If we assume that the total energy and the volume do not change, we have

$$dS = -\frac{1}{T}\sum_i \mu_i\,dn_i = -\frac{1}{T}\sum_i \mu_i \nu_i\,d\xi = \frac{\mathrm{A}\,d\xi}{T} \tag{2.29}$$

The rate of entropy change equals

$$\frac{dS}{dt} = \frac{1}{T}\mathrm{A}\frac{d\xi}{dt} \tag{2.30}$$

As follows from (2.19) and (2.15)

$$\frac{d\xi}{dt} = r^f\,(1 - e^{-\mathrm{A}/RT}) \tag{2.31}$$

and we obtain

$$\frac{dS}{dt} = r^f\,\frac{\mathrm{A}}{T}(1 - e^{-\mathrm{A}/RT}) \approx r^f\,\frac{\mathrm{A}^2}{RT^2} \quad (\text{for } \frac{\mathrm{A}}{RT} << 1) \tag{2.32}$$

The meaning of (2.32) is the following. Whatever the sign of the chemical affinity is, dS/dt is always larger than zero. When the system is close to equilibrium, the rate of entropy change is proportional to the square of the chemical affinity. As long as the entropy change is positive, the reaction proceeds.

A state in equilibrium should be stable. The deviation from equilibrium, for which the entropy is maximum, will cause a negative change in entropy, and as a consequence, the state will not change. So, expression (2.32), the condition for chemical change, is consistent with the state of equilibrium being the state of maximum entropy. Thus, the description of kinetics by an expression for the reaction rate as in (2.5) is in agreement with the second law of thermodynamics.

2.3. COUPLED REACTIONS AND THE STEADY-STATE ASSUMPTION

The steady-state approximation enables one to formulate an overall rate equation for a system of coupled reactions in terms of elementary steps. For a system at steady state, the entropy production is at minimum.

Usually a reaction is the result of a number of successive elementary steps

$$\mathrm{A + B \leftrightarrows C}$$

$$\mathrm{D + C \leftrightarrows E} \tag{2.33}$$

$$\overline{\mathrm{A + B + D \leftrightarrows E}}$$

In this example there are two elementary steps and one overall reaction equation. Let us generalize and consider a chain reaction that consists of n successive steps

$$I^{(0)} \underset{k_1^f}{\overset{k_1^b}{\leftrightarrows}} I^{(1)} \leftrightarrows \ldots \leftrightarrows I^{(i)} \leftrightarrows \ldots \underset{k_n^f}{\overset{k_n^b}{\leftrightarrows}} I^{(n)} \tag{2.34}$$

The equilibrium constants of the elementary steps i are defined as†:

---

†Of course, further generalizations are conceivable in which each elementary step involves the reaction of more than one compound and with stoichiometric coefficients larger than one, but this does not lead to kinetic expressions that are essentially different from the ones derived here.

$$K_i = \frac{k_i^f}{k_i^b} = \frac{[I^{(i)}]_{eq}}{[I^{(i-1)}]_{eq}} \tag{2.35}$$

and the overall equilibrium constant

$$K_0 = \frac{[I^{(n)}]_{eq}}{[I^{(0)}]_{eq}} \tag{2.36}$$

Substitution of (2.35) in (2.36) gives the equilibrium relation

$$K_0 = \prod_i^n K_i \tag{2.37}$$

Similarly

$$\begin{aligned} -A_0 &= \mu(I^{(n)}) - \mu(I^{(0)}) \\ &= \mu(I^{(n)}) - \mu(I^{(n-1)}) + \mu(I^{(n-1)}) - \ldots - \mu(I^{(1)}) + \mu(I^{(1)}) - \mu(I^{(0)}) \\ &= -\sum_i A_i \end{aligned} \tag{2.38}$$

We will see later that the system of coupled reactions can go to a stable state (the steady state), which lies close to equilibrium. This has a non-vanishing rate of entropy production which represents the minimum achievable under the given conditions. In general, the rate of entropy production $\sigma = dS/dt$ is the sum of the contributions from the elementary steps and becomes [see (2.30) and (2.19)]:

$$\sigma = \sum_i^n \frac{r_i A_i}{T} \tag{2.39}$$

The concentrations c_i follow from

$$\frac{dc_i}{dt} = r_i - r_{i+1} \tag{2.40}$$

and the relation between reaction rates and affinities becomes

$$r_i = k_i^f c_{i-1} - k_i^b c_i = k_i^f c_{i-1}\left(1 - K_i^{-1}\frac{c_i}{c_{i-1}}\right) = r_i^f(1 - e^{-A_i/RT}) \tag{2.41}$$

Substitution of (2.41) in (2.39) gives

$$\sigma = \sum_{i=1}^{n} \frac{r_i^f A_i(1 - e^{-A_i/RT})}{T} \approx \frac{1}{RT^2} \sum_{i=1}^{n} r_i^f A_i^2 \quad \text{for } \frac{A_i}{RT} << 1 \tag{2.42}$$

The approximation is valid if the system is close to equilibrium.

Coupled reaction equations can easily be solved using the steady-state assumption. A reaction occurs at steady state when the rate of removal of products $I^{(n)}$ equals the rate of introduction of the reactants $I^{(0)}$. When reaction conditions chosen make this true (a condition often occurring in practice), the production of intermediates $I^{(i)}$, for $0 < i < n$, is constant

$$\frac{dc_i}{dt} = 0 \quad (i = 1, 2, \ldots, n) \tag{2.43}$$

with as a consequence

$$\begin{aligned} r_i &= r_{i+1} = r_0 \quad (i = 1, \ldots, n-1) \\ r_i^f A_i &= r_0^f A_0 \quad (i = 1, \ldots, n-1) \end{aligned} \tag{2.44}$$

Expression (2.43) is the steady-state assumption. The concentrations of all intermediates are constant. They deviate from their equilibrium values, but as little as possible as we will see.

Using the expression for the entropy production, (2.42), we obtain in the steady state

$$\sigma = \frac{1}{RT^2} \sum_i r_i^f A_i^2 = \frac{1}{RT^2} r_0^f A_0 \sum_i A_i = \frac{1}{RT^2} r_0^f A_0^2 \tag{2.45}$$

As (2.44) implies

$$\sum_i A_i = \sum_i \frac{r_0^f}{r_i^f} A_0 \tag{2.46}$$

it follows with (2.38) that at steady state

$$\frac{1}{r_0^f} = \sum_i \frac{1}{r_i^f} \tag{2.47}$$

Expressions (2.45) and (2.47) can be considered as fundamental results of the steady-state approximation. The first expression relates the rate of entropy production to the chemical affinity of product and reactant. The second expression relates rates of formation of intermediates with the overall rate of product formation.

A third fundamental result valid in the steady-state approximation is that the rate of entropy production is at minimum. This follows directly from (2.45). At constant A_0

$$\left[\frac{\partial \sigma}{\partial A_i}\right]_{A_0 \text{ constant}} = 0 \quad i = 1, \ldots, n \tag{2.48}$$

and

$$\frac{\partial^2 \sigma}{\partial A_i^2} > 0 \quad i = 1, \ldots, n \tag{2.49}$$

Thus, we have the following situations: At equilibrium the system is in the state of maximum entropy; the rate of entropy production is zero. Every change in this state results in a decrease of the entropy. Therefore, equilibrium is a stable state. Steady state is a stable state of a system of coupled reactions close to equilibrium, where transformations occur close to reversibility. At steady state, the rate of entropy production is as small as possible within the boundary conditions posed to the reaction system (defined by A_0) and the system has the maximum entropy achievable in this situation.

The finding that at steady state the rate of entropy production is at minimum is very different from the situation we will encounter later (in Section 1.5) for stationary reaction systems which are far from equilibrium. Then that reaction sequence is stable in time which gives a maximum rate of entropy production. The conditions where such stable time-dependent reactions occur are irreversible conditions. Systems in which concentrations oscillate with time are an example.

Relation (2.47) relates the overall forward reaction rate r_0^f to the forward rates of the elementary reaction steps. We will illustrate its consequences for a two-step reaction:

$$I^{(0)} \underset{k_1^f}{\overset{k_1^b}{\leftrightarrows}} I^{(1)} \underset{k_2^f}{\overset{k_2^b}{\leftrightarrows}} I^{(2)} \tag{2.50}$$

or, with a more practical notation

$$R \underset{k_1^f}{\overset{k_1^b}{\leftrightarrows}} I \underset{k_2^f}{\overset{k_2^b}{\leftrightarrows}} P \tag{2.51}$$

Relation (2.47) now reduces to

$$r_0^f = \frac{r_1^f r_2^f}{r_1^f + r_2^f} \tag{2.52}$$

and the overall forward rate of production relates to the overall rate as

$$r_0 = r_0^f - r_0^b = r_0^f(1 - e^{-A_0/RT}) \tag{2.53}$$

According to (2.52) the forward rate constant depends on the concentrations of the reactant and the intermediates

$$r_0^f = \frac{k_1^f k_2^f [R][I]}{k_1^f [R] + k_2^f [I]} \tag{2.54}$$

Relation (2.54) is only valid when concentrations c_i are close to the values at equilibrium. Under this condition one may assume

$$\frac{[I]}{[R]} \approx \frac{k_1^f}{k_1^b} \tag{2.55}$$

Substitution of this equilibrium concentration of the intermediate gives

$$r_0^f = \frac{k_1^f k_2^f}{k_1^b + k_2^f}[R] \tag{2.56}$$

Within the steady-state approximation, this expression has general validity.

A reaction of the type given in (2.51) is also useful to illustrate the decrease in entropy production, when a reaction initially not in steady state decays to steady-state behavior.

The corresponding rate equation for $[I]$ is

$$\frac{d[I]}{dt} = k_1^f[R] - (k_1^b + k_2^f)[I] + k_2^b[P] \tag{2.57}$$

In steady state the rate of conversion of R is equal to the rate of production of P. We will solve equation (2.57) with the condition that $[R]$ and $[P]$ are constant. Equation (2.57) can then readily be integrated and gives for $[I]$

$$[I] = (1 - e^{-(k_1^b + k_2^f)t})\frac{k_1^f[R] + k_2^b[P]}{k_1^b + k_2^f} \tag{2.58}$$

We use as initial condition that $[I] = 0$ at t = 0.

The entropy production rate σ equals [apply Eq. (2.45)]

$$\sigma = \frac{1}{R_g T^2}\left(k_1^f [R]\, A_1^2(t) + k_2^f [I(t)]\, A_2^2(t)\right) \tag{2.59}$$

According to (2.38) we can write for A_1 and A_2

$$A_1 = \mu(I) - \mu(R)$$

$$A_2 = \mu(P) - \mu(I) \tag{2.60a}$$

Applying (2.13) for A_1, we find

$$A_1 = -RT\left(\ln\frac{I(t)}{I_{\text{eq}}} + \ln\frac{R_{\text{eq}}}{R}\right) = -RT\ln\left(\frac{R_{\text{eq}}}{I_{\text{eq}}}\cdot\frac{I(t)}{R}\right) \tag{2.60b}$$

which, after insertion of (2.3) and (2.58) yields

$$A_1 = -R_g T\ln\left[\left\{\frac{k_1^b}{k_1^f}\cdot\frac{k_1^f + k_2^b\left\{\frac{[P]}{[R]}\right\}}{k_1^b + k_2^f}\right\}\left(1 - e^{-(k_1^b + k_2^f)t}\right)\right] \tag{2.60c}$$

Similarly we find for A_2

$$A_2 = -R_g T\ln\left[\left\{\frac{k_2^b}{k_2^f}\cdot\frac{k_1^b + k_2^f}{k_1^f\frac{[R]}{[P]} + k_2^b}\right\}\left(1 - e^{-(k_1^b + k_2^f)t}\right)^{-1}\right] \tag{2.61}$$

Clearly $|A_1|$ and $|A_2|$ are infinite at t=0 and reach their minimal values at $t \to \infty$, where $[I]$ approaches its steady-state value. Figure 2.2 shows schematically how the entropy production σ decreases to its minimum value σ^{st} at steady state, which is found from insertion of (2.60) and (2.61) in (2.59) for $t \to \infty$

$$\sigma^{\text{st}} = R_g\left[k_1^f\,[R]\ln^2\left[\frac{k_1^b}{k_1^f}\,\frac{k_1^f + k_2^b\left\{\frac{[P]}{[R]}\right\}}{k_1^b + k_2^f}\right\}\right] + \\ R_g\left[k_2\frac{k_1^f[R] + k_2^b[P]}{k_1^b + k_2^f}\ln^2\left\{\frac{k_2^b}{k_2^f}\,\frac{k_1^b + k_2^f}{k_1^f\frac{[R]}{[P]} + k_2^b}\right\}\right] \tag{2.62}$$

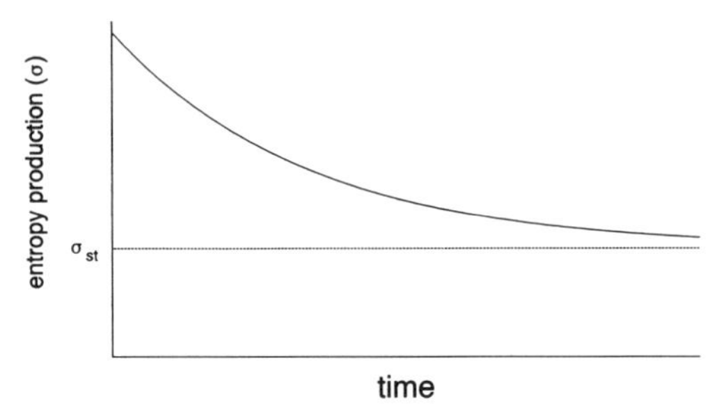

Figure 2.2. Time dependence of the entropy production σ for a reaction approaching steady-state conditions.

As discussed in Section 2.2, the entropy production should go to zero if the reaction proceeds to equilibrium. In this case

$$\frac{[P]}{[R]} = \frac{k_1^f}{k_1^b} \cdot \frac{k_2^f}{k_2^b} \tag{2.63}$$

and σ^{st} as expressed in (2.62) indeed becomes equal to zero.

An important special case of (2.56) appears if the backward reaction can be ignored ($k_2^f > k_1^b$). Now the rate of production r depends only on the rate constant k_1^f of the forward product conversion rate and one finds

$$r_0^f = \frac{d[P]}{dt} = k_1^f [R] \tag{2.64}$$

The initial conversion rate constant of the reactant R now determines the rate, and is called the *rate-determining* or *rate-limiting step.*

A different result is found when the consecutive conversion rate of the intermediate I to the product P becomes rate limiting ($k_2^f << k_1^b$). Then

$$\frac{d[P]}{dt} = k_2^f \frac{k_1^f}{k_1^b} [R] = k_2^f \, K_1^{eq} [R] \tag{2.65}$$

Because the consecutive conversion of the intermediate is slow, the reactant R and the intermediate I are in equilibrium with each other and the equilibrium constant for this step appears in the expression for the overall rate of conversion.

Note that irrespective of which of the two steps is rate determining, the rate of product formation depends linearly on the reactant concentration. Thus, one cannot distinguish between the two cases by measuring the rate $d[P]/dt$ as a function of $[R]$. As will be shown in Chapter 6, this has consequences for the interpretation of the measured activation energy as well. It is important to realize that it is generally not possible to determine the reaction mechanism and to identify the rate-determining step of a reaction sequence based on kinetic measurements alone.

Especially in catalytic processes, identification of the rate-limiting step is essential. It determines the form of the expression to be used as reaction-rate constant. From this expression, information can be deduced as to which reaction steps or surface equilibria the catalyst has to modify. We will return to this point after having analyzed the steady-state approximation further.

The steady-state condition, $d[I]/dt = 0$, also holds often for systems which are time dependent. This leads to a very useful procedure of solving kinetic equations. One can demonstrate this by applying the steady-state condition directly to the two step reaction (2.51) and will find (2.56) as the result.

The rate equation for the formation of the intermediate I is

$$\frac{d[I]}{dt} = k_1^f[R] - (k_1^b + k_2^f)[I] + k_2^b[P] \tag{2.66}$$

According to the steady-state approximation $d[I]/dt = 0$. Under the condition that the backward reaction from product P to the intermediate can be ignored, we obtain

$$[I] = \frac{k_1^f[R]}{k_1^b + k_2^f} \tag{2.67}$$

Again, if we ignore the backward reaction of P to I, the rate of production of product P follows by substitution of (2.67)

$$\frac{d[P]}{dt} = k_2^f[I] = \frac{k_1^f k_2^f}{k_1^b + k_2^f}[R] \tag{2.68}$$

This relation is the same as (2.56).

The steady-state approximation is particularly useful in flow systems with a constant input of reactant and a constant rate of product formation, as is often the case in industrial reactors.

2.4. CHAIN REACTIONS

Chain reactions provide energetically favorable reaction routes for reactions between stable molecules. The kinetics of chain reactions can be solved with the steady-state approximation.

Chain reactions form a particularly important type of coupled reactions. These were discovered around 1913, when Max Bodenstein and W. Dux found that the reaction between H_2 and Cl_2 to HCl could be brought about by irradiating the reaction mixture with photons. The very surprising result for that time was that the number of HCl molecules per absorbed photon (the quantum yield) was as high as 10^6! Although Bodenstein suggested the idea of a chain reaction in 1913, the correct explanation was given five years later by Nernst, who invoked the occurrence of a chain reaction and H and Cl radicals, in the following scheme:

$$
\begin{aligned}
Cl_2 &\stackrel{h\nu}{\rightarrow} 2\,Cl \\
Cl + H_2 &\rightarrow HCl + H \\
H + Cl_2 &\rightarrow HCl + Cl \\
2Cl &\rightarrow Cl_2
\end{aligned}
\tag{2.69}
$$

Once chlorine atoms have been produced in the first step (initiation step), the next two reactions provide a closed cycle which can be repeated many times, e.g. 10^6. This is the propagation stage of the chain process. The cycle stops when the chlorine atom recombines with another chlorine atom to form Cl_2, in what is called the termination step.

The reason why this set of coupled reactions prevails over a direct reaction between H_2 and Cl_2 becomes clear if we consider the activation energies of the reaction steps. Whereas the direct reaction between H_2 and Cl_2 requires an activation energy of over 200 kJ/mol, the activation energies of the two propagation steps are only 25 and 13 kJ/mol, respectively, which favors the chain process over the direct reaction. The difficult step is the initiation; it is facilitated by the absorption of a photon. The dynamics of elementary steps of the type $H + Cl_2 \rightarrow HCl + Cl$ have been extensively studied in the past. We will return to these reactions in Chapters 4 and 5.

We will illustrate the use of the steady-state approximation for solving the rate expressions in the reaction between hydrogen and bromine, which proceeds according to a scheme that is slightly more complicated than that for the $H_2 + Cl_2$ reaction

initiation:

$$Br_2 \underset{k_1^f}{\rightarrow} 2\,Br \tag{2.70}$$

propagation:

$$Br + H_2 \underset{k_2^f}{\overset{k_2^b}{\leftrightarrows}} HBr + H$$

$$\mathrm{H} + \mathrm{Br}_2 \underset{k_3^f}{\rightarrow} \mathrm{HBr} + \mathrm{Br} \tag{2.71}$$

termination:

$$2\mathrm{Br} \underset{k_4^f}{\rightarrow} \mathrm{Br}_2 \tag{2.72}$$

The rate of HBr formation is

$$\frac{d[\mathrm{HBr}]}{dt} = k_3^f[\mathrm{H}][\mathrm{Br}_2] + k_2^f[\mathrm{Br}][\mathrm{H}_2] - k_2^b[\mathrm{HBr}][\mathrm{H}] \tag{2.73}$$

Application of the steady-state approximation to [H]

$$\frac{d[\mathrm{H}]}{dt} = 0 = k_2^f[\mathrm{Br}][\mathrm{H}_2] - k_2^f[\mathrm{HBr}][\mathrm{H}] - k_3^f[\mathrm{H}][\mathrm{Br}_2] \tag{2.74}$$

or

$$[\mathrm{H}] = \frac{k_2^f[\mathrm{Br}][\mathrm{H}_2]}{k_3^f[\mathrm{Br}_2] + k_2^b[\mathrm{HBr}]} \tag{2.75}$$

and to [Br]

$$\frac{d[\mathrm{Br}]}{dt} = 0 = 2k_1^f[\mathrm{Br}_2] - 2k_4^f[\mathrm{Br}]^2 + k_2^b[\mathrm{H}][\mathrm{HBr}] - k_2^f[\mathrm{Br}][\mathrm{H}_2] + k_3^f[\mathrm{H}][\mathrm{Br}_2] \tag{2.76}$$

As the last three terms equal $d[\mathrm{H}]/dt$ which is zero, we obtain

$$[\mathrm{Br}] = \left(\frac{k_1^f}{k_4^f}[\mathrm{Br}_2]\right)^{1/2} \tag{2.77}$$

Replacing [H] and [Br] in the rate of HBr formation gives

$$\frac{d[\mathrm{HBr}]}{dt} = \frac{k[\mathrm{H}_2][\mathrm{Br}_2]^{1/2}}{1 + k'[\mathrm{HBr}]/[\mathrm{Br}_2]} \tag{2.78}$$

Note that the concentration-dependent term in the denominator is caused by the reversibility of the first propagation step.

Chain reactions are very common in chemistry. One well-known chain reaction is the formation of NO in car engines. Here the initiation is the thermal decomposition of an O_2 molecule at high temperature, and the thermal cracking of ethane, initiated by the decomposition of ethane in two methyl radicals and followed by a chain of reactions which produce methane and mainly ethylene. An example that has attracted much interest recently is the degradation of O_3 in the ozone layer, in the presence of chlorine

$$\begin{aligned}
&\mathrm{R-Cl} \xrightarrow{h\nu} \mathrm{R+Cl} && \text{(initiation)} \\
&\left.\begin{aligned} &\mathrm{Cl+O_3 \rightarrow O_2+ClO} \\ &\mathrm{ClO+O \rightarrow Cl+O_2} \end{aligned}\right\} && \text{(propagation)} \\
&\mathrm{RH+Cl \rightarrow R'+HCl} && \text{(termination)}
\end{aligned} \tag{2.79}$$

R-Cl is a chlorinated hydrocarbon or chlorofluorocarbon which decomposes under the influence of UV light. Oxygen atoms involved in the second propagation step are abundantly present in the higher layers of the atmosphere, where they originate from photodissociation of O_2 and NO_2. Termination of the chain occurs if Cl reacts with hydrocarbons, or other H-containing molecules to HCl.

Chain reactions bear some resemblance to catalytic reactions. The essential difference, however, is the active centers of the chain reaction appear and disappear in initiation and termination steps, whereas the number of active centers in a catalytic reaction is constant.

2.5. PARALLEL AND CONSECUTIVE REACTIONS

The rate equations of simple reactions for which the concentrations of reactants, intermediates, and products depend on time can be solved analytically. Steady-state conditions appear as a special case in the kinetics of irreversible consecutive reactions.

In this section we discuss the kinetics of simple reaction schemes that are encountered in batch processes when the concentrations of reactants, intermediates, and products depend on time. First, we consider the scheme of two irreversible, parallel reactions

$$R \xrightarrow{k_1} P_1, \qquad R \xrightarrow{k_2} P_2 \tag{2.80}$$

for which the rate equation becomes

$$-\frac{d[R]}{dt} = (k_1 + k_2)[R] \tag{2.81}$$

$$\frac{d[P_i]}{dt} = k_i\,[R]; \quad i = 1, 2 \tag{2.82}$$

Expression (2.81) is readily integrated by using the boundary condition that $[R] = [R_0]$ at $t = 0$

$$[R] = [R_0]\,e^{-(k_1+k_2)t} \tag{2.83}$$

while (2.83) leads with $[P] = 0$ at $t = 0$ to

$$[P_i] = \frac{k_i[R_0]}{k_1 + k_2}(1 - e^{-e(k_1+k_2)t}) \tag{2.84}$$

The selectivity, S, for each of the products is thus

$$S_i = \frac{k_i}{k_1 + k_2} \times 100\% \tag{2.85}$$

The case of a series of irreversible consecutive reactions is a bit more complicated.

Let us reconsider the sequence of two reaction steps from reactant R to intermediate I to product P. We will now simplify the system and assume that the reaction proceeds in one direction only:

$$R \xrightarrow{k_1^f} I \xrightarrow{k_2^f} P \tag{2.86}$$

This implies that the backward reaction-rate constants are so small that they can be ignored. The rate equations become

$$-\frac{d[R]}{dt} = k_1^f[R]$$

$$\frac{d[I]}{dt} = k_1^f[R] - k_2^f[I]$$

$$\frac{d[P]}{dt} = k_2^f[I] \tag{2.87}$$

The concentration of reactant R can readily be solved as a function of time

$$[R] = [R_0]\,e^{-k_1^f t} \tag{2.88}$$

The intermediate concentration $[I]$ is the solution of expression

$$\frac{d[I]}{dt} + k_2^f[I] = k_1^f[R_0]\,e^{-k_1^f t} \tag{2.89}$$

A standard way of solving this type of differential equation is to multiply both the left- and the right-hand side with a suitable function $f(t)$, chosen such that

$$\frac{d\,f(t)}{dt} = k_2^f f(t) \rightarrow f(t) = e^{k_2^f t} \tag{2.90}$$

Now we can write the differential equation in a form in which it is readily solved

$$\frac{d}{dt}([I]e^{k_2^f t}) = k_1^f\,[R_0]e^{(k_2^f - k_1^f)t} \tag{2.91}$$

By using $[I] = 0$ at $t = 0$ as a boundary condition, we find the following expression for $[I]$

$$[I(t)] = \frac{k_1^f}{k_2^f - k_1^f}\left(e^{-k_1^f t} - e^{-k_2^f t}\right)[R_0] \tag{2.92}$$

and, similarly, for $[P]$

$$[P(t)] = \left\{1 - \frac{k_2^f}{k_2^f - k_1^f}e^{-k_1^f t} + \frac{k_1^f}{k_2^f - k_1^f}e^{-k_2^f t}\right\}[R_0] \tag{2.93}$$

Solutions for several choices of k_1^f and k_2^f are shown in Figure 2.3. To visualize the situation, one can think of a batch reactor initially filled with reactant R, which at $t = 0$ starts to convert through the intermediate I to the product P. Note that the product formation starts with a zero slope and that it has an inflection point at the moment that the intermediate concentration reaches a maximum at $t = t_{max}$

$$t_{max} = \frac{1}{k_2^f - k_1^f}\ln\frac{k_2^f}{k_1^f}$$

$$[I]_{\max} = \left(\frac{k_1^f}{k_2^f}\right)^{\frac{k_2^f}{k_2^f - k_1^f}}[R_0] \tag{2.94}$$

There are two cases where the steady-state approximation holds. The first is for the short period around t_{max}, where $[I]$ is approximately independent of time. The second case arises when

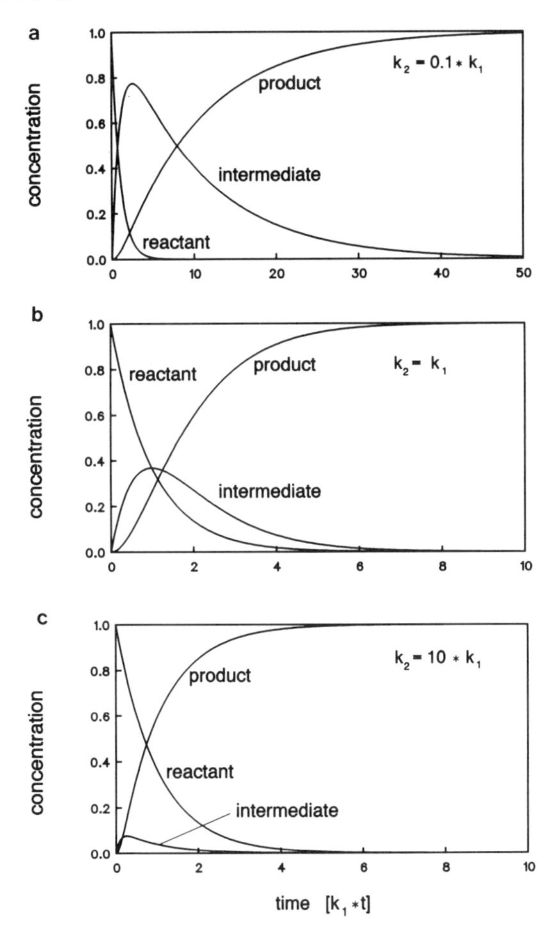

Figure 2.3. Concentration of reactant, intermediate, and product for a consecutive reaction for several choices of the rate constants.

$$t >> \frac{1}{k_2^f}; \quad k_1^f << k_2^f \tag{2.95}$$

The concentration of I remains low, because it converts to the product P as soon as it has been formed. This is the case for Figure 2.3c. Thus one has to be careful to apply the steady-state condition to systems in which concentrations vary in time, because it applies only in a limited time regime and for particular ratios of the rate constants.

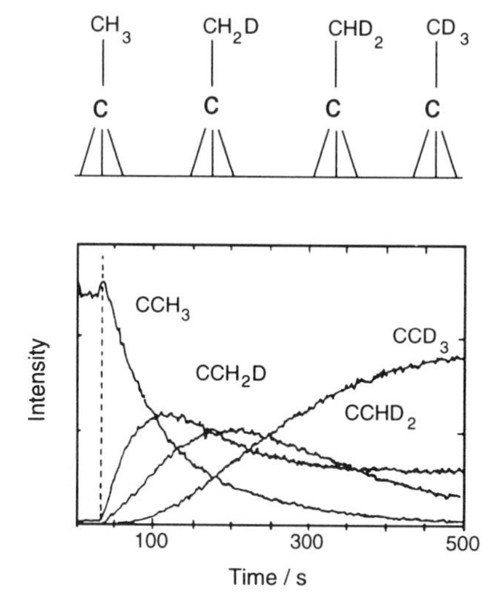

Figure 2.4. Consecutive reaction: H–D exchange in the adsorbed hydrocarbon fragment ethylidyne [adapted from J.R. Creighton, K.M. Ogle, and J.M. White, *Surf Sci.* **138**, L137 (1984)].

Figure 2.4 gives an example, taken from the work of Creighton and coworkers (1984), of a consecutive reaction between an adsorbed hydrocarbon fragment, called ethylidyne ($Pt_n{\equiv}C{-}CH_3$, where Pt_n denotes the surface of a platinum crystal), and deuterium. The reaction sequence is

$$Pt_n{\equiv}C{-}CH_3 + D \rightarrow Pt_n{\equiv}C{-}CH_2D + H$$

$$Pt_n{\equiv}C{-}CH_2D + D \rightarrow Pt_n{\equiv}C{-}CHD_2 + H \qquad (2.96)$$

$$Pt_n{\equiv}C{-}CHD_2 + D \rightarrow Pt_n{\equiv}C{-}CD_3 + H$$

The reaction proceeds under an excess of deuterium such that the backward reactions can be ignored. The concentrations of the reacting species have been followed with secondary ion mass spectrometry (SIMS). This technique yields the concentration of the reacting species through mass spectroscopic signals due to the methyl groups at atomic mass 15 (CH_3 of the reactant), 16 (CH_2D), 17 (CHD_2) and 18 (CD_3 from the fully deuterated product). Figure 2.4 shows how the concentrations vary with time. For the first 300 seconds of reaction, the pattern is exactly what one expects for a set of three consecutive reactions with two intermediates.[†]

[†]The fact that the signal due to 16 atomic mass units does not go to zero is an artifact of the SIMS technique. During measurements the species may fragment. Hence, mass 16 is not uniquely associated with the -CH_2D group of the singly deuterated ethylidyne, but can also be due to the CD_2 fragment of the triply deuterated ethylidyne species.

2.6. PRINCIPLE OF CATALYSIS

A catalyst increases the rate of a reaction by providing a reaction path with a low activation energy. Catalysts are self-regenerating systems. In a catalytic reaction, the total number of active sites is constant. When the reaction cycle is completed for a particular molecule, the surface reaction site is regenerated. The order of the reaction depends on the fraction of sites that is occupied. A maximum in the rate as a function of temperature is usually found. It indicates that the rate-limiting step changes. The conversion in a catalytic reaction can never be higher than thermodynamics predicts.

Let us reconsider the reaction between A and B to form products C and D as in (2.1) but now catalyzed by a surface as in heterogeneous catalysis, or by a catalytic complex as in homogeneous or enzymatic catalysis. Figure 2.5 illustrates how the reaction proceeds if A, B, C, and D are gas phase molecules and the catalyst is a solid surface. The essential steps in the process are the following:

1) Adsorption of the reactants on the surface, where the molecules may be adsorbed in their molecular form (CO on platinum) or be dissociated into the constituent atoms. This happens with CO on iron and H_2 on almost all metals.
2) Reaction of the adsorbed species on the surface, usually in a number of consecutive steps.
3) Desorption of the products from the surface, leaving empty sites on the surface where the following reactant molecules can adsorb.

The role of the catalytic surface is to provide an alternative reaction route for the gas-phase reaction, in which the activation energies for the separate steps are all significantly lower than for the uncatalyzed reaction. The Arrhenius rate equation

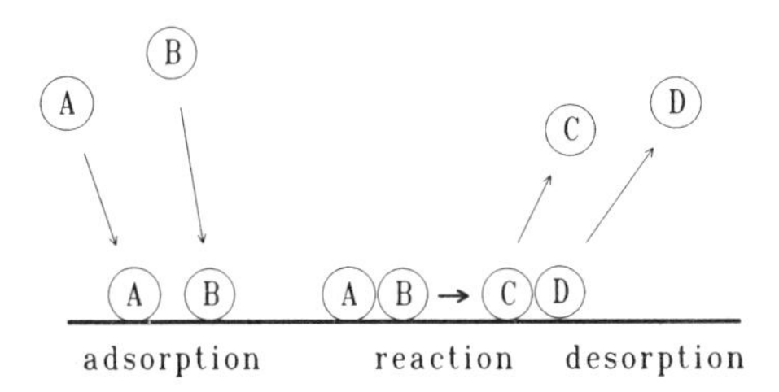

Figure 2.5. Schematic representation of a catalytic reaction of gas-phase molecules over the surface of a solid catalyst.

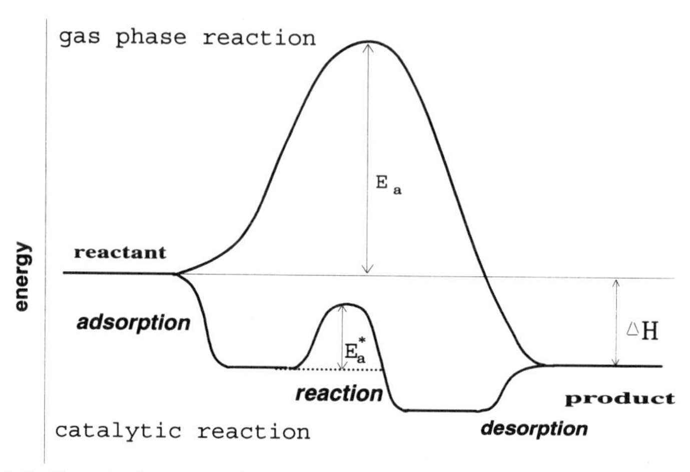

Figure 2.6. The activation energy of a catalytic reaction is lower than that of a gas phase reaction.

and the activation energy are treated in detail in Chapter 4 [see Eqs. (4.130) and (4.131)]. According to the Arrhenius rate expression, the rate increases exponentially with the temperature. In order for a reaction to occur, a reaction barrier has to be overcome. This reaction barrier is closely related to the activation energy. The overall rate of the catalytic reaction is enhanced because the overall activation energy of the catalytic reaction is lower than that of the non-catalytic reaction. This is illustrated in Figure 2.6. The adsorption energies of each of the adsorbed species involved in the catalytic reaction should not be too high, otherwise the activation barrier for the next reaction step cannot be surmounted. On the other hand, adsorption energies may not be too weak or the species would desorb from the surface and be unavailable for reaction. A discussion of the overall activation energy of a catalytic reaction and how it depends on the microscopic properties of the reacting molecules is given in Chapter 6.

Figure 2.6 illustrates another important feature of a catalytic reaction: The change in free energy before and after reaction is the same for the catalyzed and the noncatalyzed reaction. Only the activation energy is different. Thus, if thermodynamics predicts certain equilibrium concentrations under a given combination of pressure and temperature, these will also apply to the catalyzed reaction. Only the time needed to reach the equilibrium state is shorter for the catalytic reaction. In other words, a catalyst accelerates both the forward reaction from reactant to product and the backward reaction from product to reactant to the same extent. This is Ostwald's definition: *catalysis changes reaction kinetics but has no effect on thermodynamics.*

An essential property of a catalyst is that it is regenerated in its original state after each reaction cycle from reactants to products. In order to appreciate this, one should realize that a catalyst provides sites for a reactant molecule to adsorb. In its adsorbed state the molecule undergoes chemical changes and finally the product molecule desorbs, and regenerates a vacant site for adsorption of the next molecule. The overall result is a conversion of reactant to product molecules by a series of reactions in which the catalyst is first consumed but regenerated at the end. In this section we analyze the kinetics of such a self-regenerating catalytic system.

The regeneration of active sites distinguishes catalytic reactions from stoichiometric reactions. In the latter, coproducts are generated in addition to the required product. Sometimes such coproducts are desired, but often they are not and create a waste problem.

The difference between a catalytic and a stoichiometric reaction is illustrated by the selective oxidation of ethylene to ethylene epoxide, where we compare the silver-catalyzed ethylene epoxidation with the stoichiometric epichlorohydrine process. Ethylene epoxide (oxirane) has industrial importance as a starter material for the production of ethylene glycol (antifreeze) and many other products (polyethers, polyurethanes).

In the stoichiometric epichlorohydrine process, ethylene epoxide is produced in three steps:

$$\mathrm{Cl_2 + NaOH \rightarrow HOCl + NaCl}$$

$$\mathrm{C_2H_4 + HOCl \rightarrow CH_2ClCH_2OH} \tag{2.97}$$

$$\mathrm{CH_2ClCH_2OH} + \frac{1}{2}\mathrm{Ca(OH)_2} \rightarrow \frac{1}{2}\mathrm{CaCl_2 + C_2H_4O + H_2O}$$

$$\mathrm{Cl_2 + NaOH} + \frac{1}{2}\mathrm{Ca(OH)_2 + C_2H_4 \rightarrow C_2H_4O} + \frac{1}{2}\mathrm{CaCl_2 + NaCl + H_2O}$$

From the environmental point of view this is not an attractive process. In addition to the desired conversion of ethylene to ethylene epoxide, Cl_2, NaOH and $Ca(OH)_2$ are converted into $CaCl_2$ and NaCl. The traditional way of solving this waste problem has been to dump these salts in a river or the ocean, but this is no longer acceptable.

The catalytic route offers an excellent alternative for the production of ethylene epoxide. Silver is the best catalyst for this reaction. The reaction proceeds via the following elementary steps:

- dissociative adsorption of oxygen (* denotes an active site on the silver surface)

$$O_2 + 2* \rightarrow 2O_{ads} \tag{2.98}$$

- molecular adsorption of ethylene

$$C_2H_4 + * \rightarrow C_2H_{4,ads} \tag{2.99}$$

- surface reaction between adsorbed O atoms and adsorbed ethylene, followed by desorption of ethylene epoxide, in which the catalytic sites are regenerated

$$O_{ads} + C_2H_{4,ads} \rightarrow C_2H_4O + 2* \tag{2.100}$$

The overall reaction becomes

$$C_2H_4 + \frac{1}{2}O_2 \underset{Ag}{\rightarrow} C_2H_4O \tag{2.101}$$

which suggests that all reactants end up in the desired product. This is not so; the reaction produces CO_2 as a byproduct due to the total combustion of ethylene on the silver surface. However, the selectivity for ethylene epoxide can be high—around 90%. The reaction is promoted by co-adsorption of chlorine that enhances the reactivity of adsorbed oxygen with ethylene and suppresses the combustion.[†]

An essential feature of a catalytic reaction is that the active sites are regenerated. When the regeneration of the surface site is ideal, the catalyst has an infinite lifetime. In practice, after many reactant turnovers, catalytic sites may deteriorate and the catalytic activity declines. The causes for catalyst deactivation depend on the reaction. Impurities in the reactants are irreversibly adsorbed on catalytic sites. An example of this would be when lead poisons rhodium for the reaction between NO and CO to N_2 and CO_2 in the threeway catalyst for automotive pollution control. Another cause is the formation of non-selective coproducts, such as coke, which do not desorb from the catalyst. Hence, a useful catalyst enhances selectively the overall rate of the desired reaction only.

[†] The beneficial effect of chlorine was discovered inadvertently in a plant in France, when the selectivity for ethylene epoxide rose spontaneously one day. Analysis of the catalyst revealed the presence of chlorine, which originated from a neighboring chlorine plant and entered the ethylene oxide reactor together with oxygen from the air.

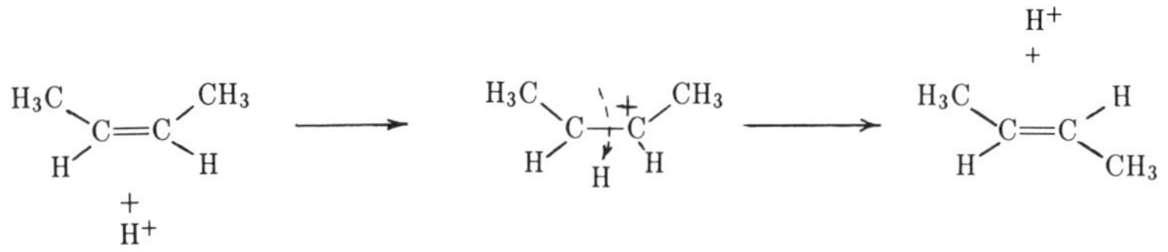

Figure 2.7. The isomerization of *cis*-butene to *trans*-butene is catalyzed by a proton which can be present as an acid site on an oxidic surface.

The kinetic description of a catalytic reaction is based on the conservation of total numbers of catalytically active sites. Let N be the total number of sites available. We will denote an empty site by an asterisk *. The reaction equations for a simple catalytic process in which a reactant R is converted to a product P via an intermediate I adsorbed on the catalyst are

$$R + * \underset{k_1^f}{\overset{k_1^b}{\leftrightarrows}} I_{\text{ads}} \tag{2.102}$$

$$I_{\text{ads}} \underset{k_2^f}{\overset{k_2^b}{\leftrightarrows}} P + * \tag{2.103}$$

An example of such a monomolecular reaction is the acid-catalyzed isomerization of *cis*-butene to *trans*-butene in Figure 2.7. The acid site is a proton as present on the surface of a solid acid (see sections 3.3 and 6.4.2.1). The adsorption complex I_{ads} formed upon protonation is a carbenium ion.

The set of equations has to be solved using explicitly the relation that conserves the total number of sites, N

$$N = n(*) + n(I_{\text{ads}}) \tag{2.104}$$

The site occupancy or surface coverage, θ, is defined as

$$\theta = \frac{n(I_{\text{ads}})}{N}; \quad 1 - \theta = \frac{n(*)}{N} \tag{2.105}$$

We will solve the set of kinetic equations for the catalytic process in (2.102) and (2.103) using the steady-state approximation. As before, we wish to compute the rate of product formation as a function of reactant or product concentration by elimination of the concentration of the intermediate from the rate expressions. The rate equations are

$$-V\frac{d[R]}{dt} = k_1^f N(1-\theta)[R] - k_1^b N\theta \qquad (2.106)$$

$$\frac{d\theta}{dt} = k_1^f(1-\theta)[R] - (k_1^b + k_2^f)\,\theta + k_2^b(1-\theta)[P] \qquad (2.107)$$

As we are now dealing with surfaces, the volume needs to be eliminated from the formulas. This, of course, has consequences for the units of the rate constants.

$$V\frac{d[P]}{dt} = k_2^f N\theta - k_2^b N(1-\theta)[P] \qquad (2.108)$$

Note that the unit of k_2^f is mol/sec.site, whereas that of k_2^b is l/sec.site.

As in the previous section, we will solve this set of equations under the assumption that the back reaction of product P does not occur. According to the steady-state approximation, $d\theta/dt = 0$ and the coverage becomes

$$\theta = \frac{\dfrac{k_1^f}{k_1^b + k_2^f}[R]}{1 + \dfrac{k_1^f}{k_1^b + k_2^f}[R]} \qquad (2.109)$$

Substituting θ in (2.108) gives us the relation between the rate of product formation and reactant concentration:

$$V\frac{d[P]}{dt} = N\frac{\dfrac{k_1^f k_2^f}{k_1^b + k_2^f}[R]}{1 + \dfrac{k_1^f}{k_1^b + k_2^f}[R]} \qquad (2.110)$$

According to Eq. (2.110) we find that the rate of product formation is proportional to the number of sites, N. In heterogeneous catalysis, reactions described by Eq. (2.110) are said to behave according to Langmuir adsorption kinetics. In homogeneous catalysis, it is called *Michaelis–Menten kinetics.*

It is interesting to compare expression (2.110) for the catalyzed reaction with its analogue (2.56) for the non-catalytic reaction. The expressions are similar, apart from the denominator in (2.110). The latter is a direct consequence of the conservation of sites (2.104). Note that the denominator can be ignored as long as the occupancy of catalytically active sites is small, $\theta << 1$. In this case the term $k_1^b + k_2^f$ is larger than k_1^f, and the reaction is first order in R. The other extreme is that k_1^f is large. That is when the rate of product formation becomes constant, or zero order in R. Thus, the reaction order can have all values between 0 and 1.

The order has a direct relation with 0, which is apparent in the following approximation for (2.110)

$$V\frac{d[P]}{dt} \approx N\,k_2^f\left(\frac{k_1^f}{k_1^b + k_2^f}[R]\right)^{(1-\theta)}$$

$$\approx N\,k_0\,[R]^{(1-\theta)} \tag{2.111}$$

expressing that the reaction may have a non-integral order. The justification for writing (2.111) is that it has the same limiting values for $\theta = 0$ or for $\theta = 1$ as (2.110) has. The approximation should only be used for coverages close to 0 or close to 1. A fractional order always indicates that the reaction mechanism consists of several elementary steps. The overall rate constant k_0 can be approximated further, depending on the rate-limiting step of the reaction.

An important special limiting case of Eq. (2.109) is adsorption in the absence of any reaction, obtained when the forward rate of reaction (2.103) is zero

$$\theta = \frac{K_1^{\text{eq}}[R]}{1 + K_1^{\text{eq}}[R]} \;;\; K_1^{\text{eq}} = \frac{k_1^f}{k_1^b} \tag{2.112}$$

which is the Langmuir isotherm. Figure 2.8 shows the coverage θ as a function of reactant concentration, $[R]$, for several values of the equilibrium constant. At low concentrations of R, θ is linear in $[R]$ and the initial slope of the curve equals

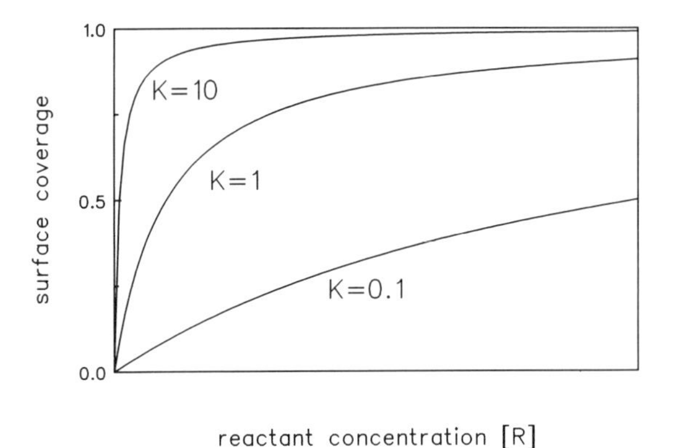

Figure 2.8. Coverage as a function of reactant concentration $[R]$ for Langmuir adsorption.

TABLE 2.1. Langmuir Isotherms for Associative, Dissociative, and Competitive Adsorption

Molecular adsorption of one type of molecule A.

$$\theta_A = K_A[A]\theta_* \; ; \; \theta_* = \frac{1}{1 + K_A[A]}$$

Competitive adsorption of two molecules A and B

$$\theta_A = K_A[A]\theta_*; \; \theta_B = K_B[B]\theta_*; \; \theta_* = \frac{1}{1 + K_A[A] + K_B[B]}$$

Dissociative adsorption of a diatomic molecule A_2

$$\theta_A = \sqrt{K_{A_2}[A_2]}\,\theta_*; \; \theta_* = \frac{1}{1 + \sqrt{K_{A_2}[A_2]}}$$

Dissociative adsorption of a diatomic molecule AB

$$\theta_A = \theta_B = \sqrt{K_{AB}[AB]}\,\theta_* \; ; \; \theta_* = \frac{1}{1 + 2\sqrt{K_{AB}[AB]}}$$

K_1^{eq} . For high values of $[R]$, θ approaches 1. Adsorption systems obey the Langmuir isotherm when interactions between adsorbed molecules can be ignored and all adsorption sites are equivalent.

Table 2.1 lists the Langmuir adsorption isotherms for several important cases, including dissociative adsorption of a diatomic molecule, and competitive adsorption of two molecules. Note that the fraction of unoccupied sites, $\theta*$, equals 1 over the denominator of the Langmuir expression. Isotherms can always be written in the form $\theta_A = K_A[A]\theta*$, which will appear very useful in solving the rate equations of complex catalytic mechanisms.

As an example, we derive the rate equation for a catalytic reaction between A and B. In this case, the case that the surface reaction between A_{ads} and B_{ads} determines the rate and proceeds in the forward direction only, while the product AB desorbs instantaneously without readsorption. The elementary reactions are

$$\begin{aligned}
A + * &\leftrightarrows A_{\text{ads}} && (1)\ \textit{equilibrium} \\
B + * &\leftrightarrows B_{\text{ads}} && (2)\ \textit{equilibrium} \\
A_{\text{ads}} + B_{\text{ads}} &\rightarrow AB_{\text{ads}} + * && (3)\ \textit{rds} \\
AB_{\text{ads}} &\rightarrow AB + * && (4)\ \textit{fast}
\end{aligned} \tag{2.113}$$

Because the product desorbs immediately upon formation, A_{ads} and B_{ads} are the only species on the surface with appreciable coverages, given by the Langmuir expression

$$\theta_A = K_A[A]\theta_*; \quad \theta_B = K_B[B]\theta_*; \quad \theta_* = \frac{1}{1 + K_A[A] + K_B[B]} \tag{2.114}$$

The rate of AB production equals

$$V\frac{d[AB]}{dt} = Nk_4\,\theta_{AB} \tag{2.115}$$

Application of the steady-state approximation to the intermediate AB_{ads} gives

$$\frac{d\theta_{AB}}{dt} = k_3\,\theta_A\theta_B - k_4\theta_{AB} = 0 \rightarrow k_4\theta_{AB} = k_3\theta_A\theta_B \tag{2.116}$$

and thus the rate of reaction equals that of the rate-determining step and becomes

$$V\frac{d[AB]}{dt} = Nk_3\theta_A\theta_B$$

$$= \frac{Nk_3K_AK_B[A][B]}{(1 + K_A[A] + K_B[B])^2} \tag{2.117}$$

Limiting cases arise if one reactant adsorbs more strongly to the catalyst than the other or when both reactants interact weakly with the surface. Suppose that B forms a much stronger bond with the catalyst than A does, such that $K_B[B] >> K_A[A]$ and $K_B[B] >> 1$.

Then we can write (2.117) as

$$V\frac{d[AB]}{dt} = N\frac{k_3K_A}{K_B}\frac{[A]}{[B]} \tag{2.118}$$

and the reaction has a positive first order in $[A]$ and a negative first order in $[B]$. The species B_{ads} has a high surface coverage and is called the majority reacting intermediate, sometimes abbreviated as *mari*. If, on the other hand, the gases adsorb only weakly, such that both $K_A[A] << 1$ and $K_B << 1$, the rate exhibits a positive first order in both reactants

$$V\frac{d[AB]}{dt} = Nk_3K_AK_B[A][B] \tag{2.119}$$

The surface of the catalyst is practically empty, implying that the active sites themselves form the majority reacting intermediate.

Figure 2.9 illustrates the expressions (2.114) and (2.117) for the case that B adsorbs more strongly than A (heats of adsorption of 100 and 94 kJ/mol, respec-

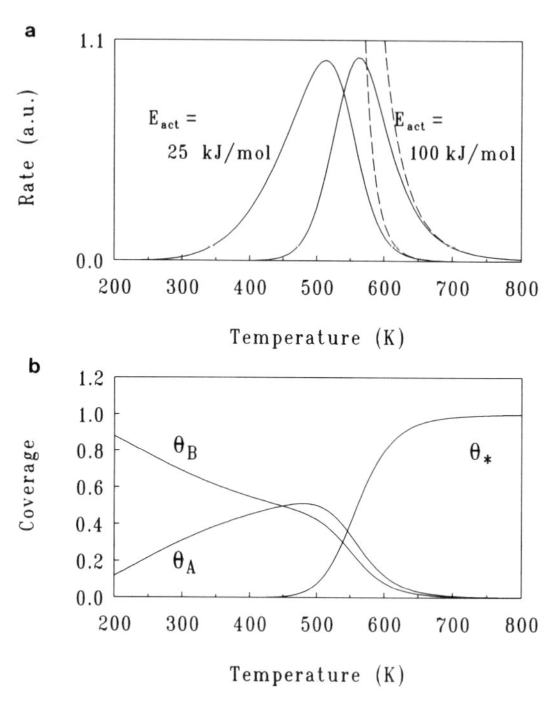

Figure 2.9. Simulation based on mechanism (2.113) for the catalytic reaction between A and B, with competitive equilibrium adsorption of A and B, a rate-limiting irreversible surface reaction between A_{ads} and B_{ads}, and a kinetically insignificant irreversible desorption of the product AB. Dashed lines represent the high temperature limits. Input values: $\Delta H_A = -94$kJ/mol, $\Delta H_B = -100$kJ/mol, $K_A[A] = 10^{-9}$, $K_B[B] = 2 \cdot 10^{-10}$; the reaction rates have been scaled to unity at maximum.

tively). In order to compensate for the difference in adsorption bond strength, A has been given a concentration five times higher than B. Figure 2.9a shows what the effect is: B_{ads} dominates the surface at low temperature, but A_{ads} acquires a comparable coverage as B at higher temperatures, owing to its higher gas phase concentration. The rates of reaction corresponding to two different activation energies of the rate-determining step are given in Figure 2.9b. Both rates have been scaled such that their maxima equal unity. The important point to note is that, whereas the rate of the homogeneous gas phase reaction between A and B increases monotonically and steeply with increasing temperature, the overall rate of the catalyzed reaction reaches its maximum and returns to zero at very high temperatures. Under such conditions, the surface coverages of A and B become zero. Also included in Figure 2.9b is the high temperature limit of the reaction rate, (2.119), which applies only if the surface is almost empty.

In general, catalytic reaction mechanisms are considerably more complex than the scheme of (2.113). Reaction intermediates other than adsorbed reactants or products may appear, complicating the expressions for the coverages of all species present on the surface. Moreover, all reaction steps may in principle proceed both in the forward and the reverse direction. The kinetics of such mechanisms is usually treated under the assumption that one of the elementary steps determines the rate while all other steps are essentially at equilibrium. We illustrate the use of this approach by discussing the kinetics of ethylene hydrogenation.

The hydrogenation of ethylene is catalyzed by many metals; orders of the reaction in ethylene and hydrogen are given in Table 2.2. We assume the following reaction mechanism, based on molecular adsorption of ethylene, dissociative adsorption of hydrogen, and the stepwise addition of hydrogen atoms to ethylene on the surface of the catalyst

$$
\begin{aligned}
C_2H_4 + * &\leftrightarrows C_2H_{4,ads} && (1)\\
H_2 + 2* &\leftrightarrows 2H_{ads} && (2)\\
C_2H_{4,ads} + H_{ads} &\leftrightarrows C_2H_{5,ads} + * && (3)\\
C_2H_{5,ads} + H_{ads} &\leftrightarrows C_2H_{6,ads} + * && (4)\\
C_2H_{6,ads} &\leftrightarrows C_2H_6 && (5)
\end{aligned}
\qquad (2.120)
$$

TABLE 2.2. Reaction Orders in H_2 and C_2H_4 for Ethylene Hydrogenation over Transition-Metal Catalysts[a]

Metal	T(K)	Order in H_2	Order in C_2H_4
Rh	197	0.85	–0.79
Ru	203	0.95	–0.59
Co	213	0.56	–0.19
Ni	—	0.67–1.0	–0.6–0
Pt	323	0.5–1.3	–0.8–0
Pd	243	0.66	–0.03
Fe	303	0.69	–0.03
Cu	353	0.69	+0.06

[a]Data from Schuit and van Reijen (1958) and Zaera and Somorjai (1984).

The rate of reaction is that of the rate-determining step, which is either reaction (3) or (4). We investigate both cases.

I) If we take the addition of the first H atom, (reaction (3)) as the rate-determining step and all other steps at equilibrium, then

$$r = V\frac{d[C_2H_6]}{dt} = Nk_3^f\,\theta_{C_2H_4}\,\theta_H - N\,k_3^b\theta_{C_2H_5}\theta_* \tag{2.121}$$

The coverages of the species in equilibrium with the gas phase—ethylene, hydrogen, and ethane—follow from the Langmuir adsorption isotherms in which the fraction of free sites is not yet known

$$\theta_{C_2H_4} = K_1[C_2H_4]\theta_*$$

$$\theta_H = K_2^{1/2}[H_2]^{1/2}\theta_*$$

$$\theta_{C_2H_6} = K_5[C_2H_6]\theta_* \quad (K_5 = k_{ads}/k_{des}) \tag{2.122}$$

The coverage of $C_2H_{5,ads}$ is determined by the equilibrium step (4)

$$k_4^f\,\theta_{C_2H_5}\,\theta_H = k_4^b\,\theta_{C_2H_6}\,\theta_* \tag{2.123}$$

or

$$\theta_{C_2H_5} = K_2^{-1/2}K_4^{-1}K_5[C_2H_6][H_2]^{-1/2}\theta_* \tag{2.124}$$

The fraction of unoccupied sites follows from the condition that the total number of sites must be conserved

$$\theta_* + \theta_{C_2H_4} + \theta_H + \theta_{C_2H_5} + \theta_{C_2H_6} = 1 \tag{2.125}$$

which yields

$$\theta_* = \frac{1}{1 + K_1[C_2H_4] + K_2^{1/2}[H_2]^{1/2} + K_2^{-1/2}K_4^{-1}K_5[C_2H_6][H_2]^{-1/2} + K_5[C_2H_6]} \tag{2.126}$$

Thus, the rate becomes

$$V\frac{d[C_2H_6]}{dt} =$$

$$\frac{Nk_3^f(K_1K_2^{1/2}[C_2H_4][H_2]^{1/2} - K_3^{-1}K_2^{-1/2}K_4^{-1}K_5[C_2H_6][H_2]^{-1/2})}{(1 + K_1[C_2H_4] + K_2^{1/2}[H_2]^{1/2} + K_2^{-1/2}K_4^{-1}K_5[C_2H_6][H_2]^{-1/2} + K_5[C_2H_6])^2} \tag{2.127}$$

We simplify the latter expression by assuming that the back reaction may be ignored and that the coverage of species, other than adsorbed ethylene, is small. This gives

$$V\frac{d[C_2H_6]}{dt}=N\frac{k_3^f\,K_1K_2^{1/2}[C_2H_4][H_2]^{1/2}}{(1+K_1[C_2H_4])^2} \qquad (2.128)$$

as the rate of reaction for our Mechanism I.

II) If we take the addition of the second H atom as the rate-determining step, then the rate becomes

$$r=V\frac{d[C_2H_6]}{dt}=Nk_4^f\,\theta_{C_2H_5}\theta_H-Nk_4^b\theta_{C_2H_6}\theta_* \qquad (2.129)$$

The coverage of $C_2H_{5,ads}$ now follows from equilibrium step (3)

$$k_3^f\,\theta_{C_2H_4}\,\theta_H=k_3^b\theta_{C_2H_5}\,\theta_* \qquad (2.130)$$

or

$$\theta_{C_2H_5}=K_1K_2^{1/2}K_3[C_2H_4][H_2]^{1/2}\theta_* \qquad (2.131)$$

The fraction of unoccupied sites becomes

$$\theta_*=\frac{1}{1+K_1[C_2H_4]+K_2^{1/2}[H_2]^{1/2}+K_1K_2^{1/2}K_3[C_2H_4][H_2]^{1/2}+K_5[C_2H_6]} \qquad (2.132)$$

and the rate

$$V\frac{d[C_2H_6]}{dt}=$$

$$\frac{Nk_4^f(K_1K_2K_3[C_2H_4][H_2]-K_4^{-1}K_5[C_2H_6])}{(1+K_1[C_2H_4]+K_2^{1/2}[H_2]^{1/2}+K_1K_2^{1/2}K_3[C_2H_4][H_2]^{1/2}+K_5[C_2H_6])^2} \qquad (2.133)$$

Again, we assume that the back reaction does not occur, and that ethylene and the intermediate $C_2H_{5,ads}$ are the only species with appreciable coverages, which leads to

$$V\frac{d[C_2H_6]}{dt}=N\frac{k_4^fK_1K_2K_3[C_2H_4][H_2]}{(1+K_1[C_2H_4](1+K_2^{1/2}K_3[H_2]^{1/2}))^2} \qquad (2.134)$$

as the rate for Mechanism II.

TABLE 2.3. Orders in Ethylene and Hydrogen for Mechanisms I and II

	Mechanism I		Mechanism II	
Temperature	H_2	C_2H_4	H_2	C_2H_4
Low	1/2	–1	$0 < x < 1$	–1
Intermediate	1/2	$-1 < y < 1$	$0 < x < 1$	$-1 < y < 1$
High	1/2	1	1	1

To test our mechanisms, we compare the rate expressions with a power rate law of the form

$$r = k_{\text{eff}}[H_2]^x[C_2H_4]^y \tag{2.135}$$

for which the orders are given in Table 2.2. The range of orders that the two mechanisms can account for is given in Table 2.3. In both mechanisms the surface will be largely occupied at low temperatures and mainly empty at high temperatures. We conclude that the reaction mechanism, in which the second addition of an H atom determines the rate, would be consistent with the experimental data in Table 2.2. Note that this does not prove that the mechanism is necessarily correct, it is certainly possible to devise other mechanisms that are consistent with the data of Table 2.2. Mechanism I, however, should be rejected, because it cannot explain orders of hydrogen in excess of 1/2.

The kinetics of the ethylene hydrogenation shows a similar temperature–dependent behavior as the simulation of Figure 2.9a. The rate of reaction increases initially with temperature, goes through a maximum and decreases thereafter. According to Zur Strassen, who investigated the ethylene hydrogenation over nickel catalysts, the reaction is close to first order in hydrogen at temperatures below the inversion point of about 170°C, but close to zero order in ethylene. However, at temperatures well above 170°C the reaction becomes first order in ethylene.

The reason for this change in kinetics lies in the surface concentrations of ethylene and hydrogen atoms. At low temperatures, the surface is almost completely covered by ethylene. The overall rate of hydrogenation is slow because pairs of free surface sites necessary for the dissociation of hydrogen are scarce and have to be created by the removal of ethylene molecules. Yet, when the temperature increases, the equilibrium concentration of adsorbed ethylene decreases, because the equilibrium constant K_1 decreases. As a result, sites for hydrogen adsorption and dissociation become available and the reaction may proceed. At high temperatures both hydrogen and ethylene are present in low concentrations and a higher gas phase concentration helps to enhance their surface coverages and consequently the rate of the reaction. This behavior is clearly reflected in the limiting cases of the rate expression (2.134)

$$V\frac{d[C_2H_6]}{dt}=N\frac{k_4^f\,K_2K_3[H_2]}{K_1[C_2H_4](1+K_2^{1/2}K_3[H_2]^{1/2})^2}\quad \text{(low } T\text{)} \qquad (2.136)$$

$$V\frac{d[C_2H_6]}{dt}=N\,k_4^f\,K_1K_2K_3[C_2H_4][H_2]\quad \text{(high } T\text{)} \qquad (2.137)$$

The rate constant as expressed by the low temperature limit increases with temperature because the equilibrium constant for ethylene adsorption, K_1, decreases, paralleled by the creation of empty sites on the surface. The decrease in rate with increasing temperature predicted by the high temperature limit is due to the decrease of the equilibrium constants K_1 and K_2, which leads to decreasing surface coverages of ethylene and hydrogen.

To determine the optimum conditions for a reaction, one requires information on its equilibrium constant. An exothermic reaction occurs at low temperature, whereas an endothermic reaction has an enhanced equilibrium favorable for product formation at high temperatures. Ethylene hydrogenation is an exothermic process. A decrease in maximum conversion is expected with increasing temperature. The increase in rate constant at temperatures below the inversion temperature found by Zur Strassen implies that conversion is far from equilibrium and has a low conversion. When the temperature of a reaction becomes too low, the values of the rate constant decrease, prohibiting the exothermic reactions.

It should be noted that several other mechanisms may be devised to explain the kinetics of the ethylene hydrogenation. For example, Cartright and coworkers explained their data with a kinetic mechanism in which ethylene and hydrogen adsorb non-competitively at low temperature on a surface mainly covered by hydrocarbon species (the Horiuti – Polanyi mechanism). At a higher temperature, however, ethylene and hydrogen start to compete for the same sites. Interestingly, these authors find that the kinetic order in hydrogen increases continuously from half order at ~250 K to first order at 340 K, which is in qualitative agreement with the rate expression (2.134).

Endothermic reactions can proceed at higher temperatures. At very high temperatures such reactions become also possible as non-catalyzed reactions and proceed by radical reactions in the gas phase. An important endothermal process is steam reforming over nickel catalysts ($CH_4 + H_2O \rightarrow CO + 3H_2$). An example of a non-catalytic process is pyrolysis ($C_2H_4 \rightarrow C_2H_2 + H_2$). The latter reaction occurs at ~1500°C, where catalysis no longer plays a role.

A general feature of heterogeneous catalytic reactions is that dissociation of molecular bonds occurs. Product formation is the result of an association reaction in which adsorbed fragments recombine. In the earlier example of selective oxidation, molecular oxygen had to dissociate; in the case of ethylene hydrogenation, dissociation of hydrogen is necessary. In ammonia synthesis, both reactants have to be dissociated. We discuss the kinetics of dissociative adsorption in more detail in Chapter 6.

2.7. STEADY STATES FAR FROM EQUILIBRIUM; AUTOCATALYSIS

Conditions forcing a system to be far removed from the state of equilibrium may give rise to self organization on a supra-molecular scale. Such systems have a low entropy, but a high rate of entropy production.

A reaction at steady state close to equilibrium has a minimum rate of entropy production. As long as the deviation from equilibrium is small, the system will always move to and not away from the equilibrium state. Prigogine (1980) has shown that the law of minimum rate of entropy production is by no means a general law. Stationary states are not confined to close-to-equilibrium conditions, but may develop far from equilibrium as well. Such stationary states are accompanied by a maximum rate of entropy production. This occurs in autocatalytic systems. Under special conditions, such systems may form stationary states where the product concentration oscillates.

Before discussing oscillating reactions, it is useful to know the roles entropy and entropy production play in the time evolution of reacting systems. In equilibrium thermodynamics we deal with a closed system of reacting molecules. No matter what the initial concentrations, the system moves to the state of equilibrium. The concentrations of all molecules participating in the reaction satisfy the law of mass action of Guldberg and Waage (2.3), with an equilibrium constant set by the free energies of the molecules, through van't Hoff's relation (2.4). The potential function of equilibrium thermodynamics is the free energy, which strives for a minimum value. At equilibrium, the entropy assumes its largest possible value, implying maximum disorder. As to evolution, the final state of the system is entirely determined by general laws and has lost all information about the initial state: Once at equilibrium, the system has no memory about its origin.

Kinetics belongs to the domain of non-equilibrium thermodynamics. If the deviation from the state of equilibrium is relatively small, linearized non-equilibrium thermodynamics applies. An example in physics is the laminar flow of a liquid through a tube under the influence of a small pressure gradient. The outstanding example of a chemical system described by linear thermodynamics is a flow reactor working under steady-state conditions. This is an open system continuously fed with reactants. The potential function which describes the reaction system is the entropy production, $\sigma = dS/dt$. As derived in Section 2.3, systems which are forced to deviate from equilibrium move toward the state of minimum entropy production. The continuous supply of reactants lowers the entropy of the system and precludes its evolution toward equilibrium. A steady state sets in at the smallest deviation from equilibrium that is possible under the given conditions. Additionally, in linearized non-equilibrium thermodynamics the evolution of a system is completely predictable and proceeds to a state dictated by general laws, without any memory for events in the past.

If the deviation from equilibrium becomes larger than a certain threshold value, the system may become unstable and exhibit chaotic or oscillatory behavior. A well-known example from physics is the transition from laminar to turbulent flow of a liquid through a pipe, if the pressure gradient exceeds a critical value corresponding to the Reynolds number. In chemistry, reactions may start to oscillate, provided the mechanism contains an autocatalytic step. Such phenomena tend to be associated with disorder, but this is not correct. For instance, turbulent flow shows an extremely high degree of organization, in which incredibly large numbers of molecules move coherently through circular patterns of macroscopic dimensions. Conditions far from equilibrium may give rise to self organization on a supramolecular scale. General characteristics of systems far from equilibrium include low entropy, high entropy production, and deviation from equilibrium larger than a critical threshold value. The system requires a continuous supply of either energy or matter, to keep the entropy low and to maintain its large deviation from the state of equilibrium.

2.7.1. Autocatalysis and Oscillating Reactions

In autocatalytic reactions, reactants catalyze their own production. This can lead to oscillating behavior in reacting systems that are far from equilibrium.

Self organization occurs in chemistry as well, where it may lead to oscillations if the reaction mechanism contains one or more autocatalytic steps:

$$X + Y \rightarrow 2X \tag{2.138}$$

The reaction is called autocatalytic, because X catalyzes its own production. Autocatalytic reactions occur frequently in biological systems. Erwin Schrödinger has made the observation that living matter evades decay to equilibrium, the state of maximum entropy. If a state is outside equilibrium, there has to be steady compensation for the production of entropy to prevent the system from falling back to its equilibrium state. This implies that the system has to be fed by free energy or by energy-rich matter. Dissipative systems, systems which exchange energy with its environment, tend to organize themselves in a stationary state far from equilibrium.

The glycolytic cycle is important for the energetics of living cells. Glucose degrades via several catalytic enzymatic steps, in which two molecules of adenosine triphosphate (ATP) are consumed. During the cycle, however, four ATP molecules are liberated. The net result is that one molecule of glucose is consumed and two molecules of ATP are produced. However, to produce ATP, ATP is needed. The ATP molecule acts as the store and supplier of energy in the cells of living organisms.

The eight elementary steps that generate energy by oxidizing glucose to pyruvic acid are:

$$\text{glucose} + \text{ATP} \rightarrow \text{glucose} - 6 - \text{phosphate} + \text{ADP} \qquad (a)$$

$$\text{glucose} - 6 - \text{phosphate} \leftrightarrows \text{fructose} - 6 - \text{phosphate} \qquad (b)$$

$$\text{fructose} - 1:6 - \text{phosphate} + \text{ATP} \rightarrow \text{fructose} - 1:6 - \text{diphosphate} + \text{ADP} \qquad (c)$$

$$\text{fructose} - 1:6 - \text{diphosphate} \leftrightarrows 2\ 3 - \text{phosphoglyceraldehyde} \qquad (d)$$

$$2\ 3 - \text{phosphoglyceraldehyde} + 2\ \text{DPN} + 2\ \text{ADP} + 2\ \text{POH} \leftrightarrows$$
$$2\ 3 - \text{phosphoglyceric acid} + 2\ \text{DPNH}_2 + 2\ \text{ATP} + 2\ \text{H}_2\text{O} \qquad (e)$$

$$2\ 3 - \text{phosphoglyceric acid} \leftrightarrows 2\ 2 - \text{phosphoglyceric acid} \qquad (f)$$

$$2\ 2 - \text{phosphoglyceric acid} \leftrightarrows 2\ \text{phosphoenolpyruvic acid} \qquad (g)$$

$$2\ \text{phosphoenolpyruvic acid} + 2\ \text{ADP} \leftrightarrows 2\ \text{pyruvic acid} + 2\ \text{ATP} \qquad (h)$$

The overall reaction is

$$\text{glucose} + 2\ \text{DPN} + 2\ \text{ADP} + 2\ \text{POH} \rightarrow$$
$$2\ \text{pyruvic acid} + 2\ \text{DPNH}_2 + 2\ \text{ATP} + 2\ \text{H}_2\text{O} \qquad (i)$$

Two molecules ATP are consumed in elementary steps (*a*) and (*c*), while four molecules ATP are formed, two in step (*e*) and two in step (*h*). The general form of the equation is

$$2\ \text{ATP} + X \rightarrow \text{ATP} + Y \qquad (j)$$

which emphasizes the autocatalytic nature of the process: ATP catalyzes its own production.

We will now discuss the entropy and entropy production of systems far from equilibrium. We expand the entropy of a system around its stationary value in a Taylor expansion as

$$S = \bar{S} + \delta S + \frac{1}{2}\delta^2 S \tag{2.139}$$

At equilibrium, S is a maximum. Therefore, δS equals zero and $\delta^2 S$ is negative. This is demonstrated by explicit calculation of $\delta^2 S$ at the equilibrium value of the entropy, $\bar{S} = S_{eq}$:

$$\delta^2 S = \sum_{ij} \frac{\partial^2 S}{\partial n_i \partial n_j} \delta n_i \delta n_j = -\frac{1}{T}\sum_{ij} \frac{\partial \mu_i}{\partial n_j} \delta n_i \delta n_j < 0 \tag{2.140}$$

To deduce (2.140) we have used (2.28) in which E and V are constant. The expression for the excess rate of entropy production is

$$\frac{1}{2}\frac{\partial}{\partial t}\delta^2 S = -\frac{1}{T}\sum_{ij} \frac{\partial \mu_i}{\partial n_j} \delta \dot{n}_i \delta n_j \tag{2.141}$$

Close to equilibrium (2.16) applies

$$\delta \dot{n}_i = -\frac{1}{RT} r_i^f \delta A_i \tag{2.142}$$

and we find

$$\frac{1}{2}\frac{\partial}{\partial t}\delta^2 S = \frac{1}{RT^2}\sum_{i} r_i^f \delta A_i^2 > 0 \tag{2.143}$$

The result (2.143) is the same as (2.49). It states that the rate of excess entropy production is positive for a stationary state close to equilibrium. At steady state, the rate of entropy production is minimum.

Let us now calculate the excess entropy production for the autocatalytic reaction. The chemical reaction affinity becomes

$$-A = \Delta\mu^0 + RT \ln \frac{[X]^2}{[X][Y]} \tag{2.144}$$

For constant $[Y]$, the variation in A becomes

$$\delta A = RT\frac{\delta[X]}{[X]} \tag{2.145}$$

Far from equilibrium, the backward rate r^b can be ignored, so that

$$\delta r = k^f [Y]\delta[X] \tag{2.146}$$

and

$$\frac{d}{dt}\delta^2 S = \frac{1}{T}\delta r\, \delta A = -Rk^f\frac{[Y]}{[X]}(\delta[X])^2 < 0 \tag{2.147}$$

Far from equilibrium, the excess rate of entropy production may become negative and the system may become unstable. It will not converge to a steady state because the excess rate of entropy production will not become a minimum. Oscillations of the reaction may be the result.

A system of coupled autocatalytic reactions that results in stable oscillations is given by

$$A + X \xrightarrow{k_1} 2X$$

$$X + Y \xrightarrow{k_2} 2Y$$

$$Y \xrightarrow{k_3} E \tag{2.148}$$

This system of coupled autocatalytic reactions was first studied by Lotka and Volterra. Once X is formed, a second species Y appears. The reactions are autocatalytic in X as well as in Y.

Interestingly, this set of reactions can be (and has been) used in the modelling of ecological systems. Imagine for example, that A represents grass, X is a deer living on A, and Y is a lion that can only survive on X. This ecosystem shows oscillations in the populations of deer and lions.

We will solve the kinetic expressions for reaction (2.148) with constant concentrations of reactant A and product E

$$\begin{aligned} \frac{1}{2}\frac{d[X]}{dt} &= k_1[A][X] - k_2[X][Y] \\ \frac{1}{2}\frac{d[Y]}{dt} &= k_2[X][\mathrm{Y}] - k_3[Y] \end{aligned} \tag{2.149}$$

The autocatalytic properties are immediately apparent from (2.149). The increase in the concentrations of X and Y depends on the concentrations of X and Y. Lotka and Volterra have shown that the system (2.148) can indeed develop stable oscillations around the steady-state solutions, which are

$$[X_0] = \frac{k_3}{k_2}; \quad [Y_0] = \frac{k_1}{k_2}[A] \tag{2.150}$$

Because we anticipate oscillations, we write the solutions of [X] and [Y] as the sum of the steady-state concentrations and a relatively small time-dependent perturbation

$$[X](t) = [X_0] + x\,e^{\omega t}; \quad [Y](t) = [Y_0] + y\,e^{\omega t} \tag{2.151}$$

with $x << [X_0]$ and $y << [Y_0]$ and we investigate under which conditions (2.151) becomes a solution of (2.149). Substitution of (2.151) into (2.149), maintaining only terms linear in x or y, gives

$$\omega x = -k_3 y; \quad \omega y = k_1[A]x \tag{2.152}$$

One then obtains as dispersion relation

$$\omega^2 + k_1 k_3[A] = 0 \rightarrow \omega = \pm\, i\sqrt{k_1 k_3[A]} \tag{2.153}$$

The time-dependent perturbation of (2.151) becomes complex and because

$$e^{i\omega t} = \cos\omega t + i \sin\omega t \tag{2.154}$$

this means that the concentrations of X and Y oscillate in time, as sketched in Fig. 2.11. The system rotates around its steady-state value, with a frequency that depends on the reactant concentration, $[A]$. We recognize the trends that we intuitively expected for the ecosystem mentioned above: the lion population (Y), starts to grow when the deer population (X) is high, but the latter decreases rapidly as there are too many lions. The lions die and the deer population starts to grow.

Other systems exist that exhibit stable cyclic behavior. If disturbed, the system will change until it has reached a cyclic state that remains stable. Essential for the occurrence of oscillations is that the mechanism contains autocatalytic steps and that there is a continuous input of reactant, which keeps the entropy low and the reaction system far from equilibrium.

A classical example of an inorganic set of coupled reactions, some of them being autocatalytic, occurs in the Belousov–Zhabotinsky reaction. This reaction, discovered in 1958 by B.P. Belousov and reported to the western world by A.N. Zaikin and A.M. Zhabotinsky in 1970, is the most investigated and best understood

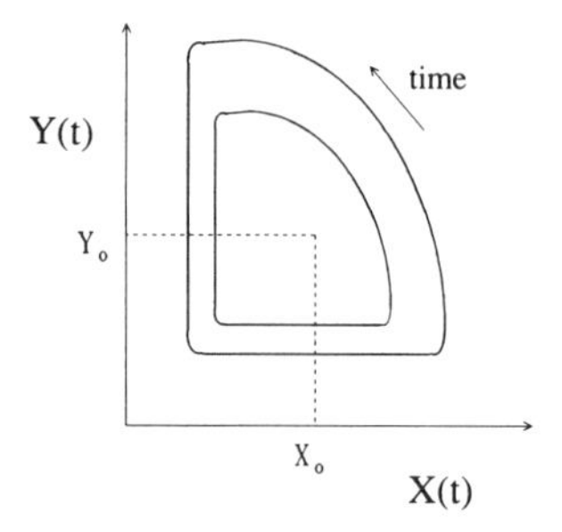

Figure 2.10. Periodic solutions of the Lotka–Volterra reactions (2.148) for different initial conditions.

example of an oscillating system of chemical reactions. The ingredients of the reaction system are malonic acid, $CH_2(COOH)_2$; bromate, BrO_3^- ; bromide, Br^-; bromous acid, $HBrO_2$; bromomalonic acid, $BrCH(COOH)_2$; and cerium ions, Ce^{3+} and Ce^{4+}. When brought together in the appropriate concentrations in an acidic solution, oscillations develop which are easily monitored by following the concentrations of the bromide and cerium ions (Figure 2.10). The frequency of the oscillations depend on the initial concentrations of malonic acid, bromate, and cerium.

The detailed reaction mechanism, as described by Noyes and coworkers (1990), is very complicated. We follow a somewhat simpler description given by Pacault *et al.* (1976) in three processes.

Process A is the reaction between bromide and bromate to bromine, which reacts with malonic acid to bromomalonic acid.

$$BrO^{3-} + Br^- + 2H^+ \rightleftarrows HOBr + HBrO_2 \tag{A-1}$$

$$HBrO_2 + Br^- + H^+ \rightleftarrows 2HOBr \tag{A-2}$$

$$3\{HOBr + Br^- + H^+ \rightleftarrows Br_2 + H_2O\} \tag{A-3}$$

$$3\{Br_2 + CH_2(COOH)_2 \rightarrow HBr^- + BrCH(COOH)_2\} \tag{A-4}$$

or, in total

$$BrO_3 + 2Br^- + 3CH_2(COOH)_2 + 3H^+ \rightarrow 3BrCH(COOH)_2 + 3H_2O \tag{A-5}$$

Process B is the autocatalytic formation of bromous acid with a simultaneous and rapid oxidation of Ce^{3+} to Ce^{4+}

$$BrO_3^- + HBrO_2 + H^+ \rightleftarrows 2BrO_2 + H_2O \tag{B-1a}$$

$$2\{BrO_2 + Ce^{3+} + H^+ \rightleftarrows Ce^{4+} + HBrO_2\} \tag{B-1b}$$

These two reactions add up to

$$2Ce^{3+} + HBrO_2 + BrO_3^- + 3H^+ \rightarrow 2Ce^{4+} + 2HBrO_2 + H_2O \tag{B-1}$$

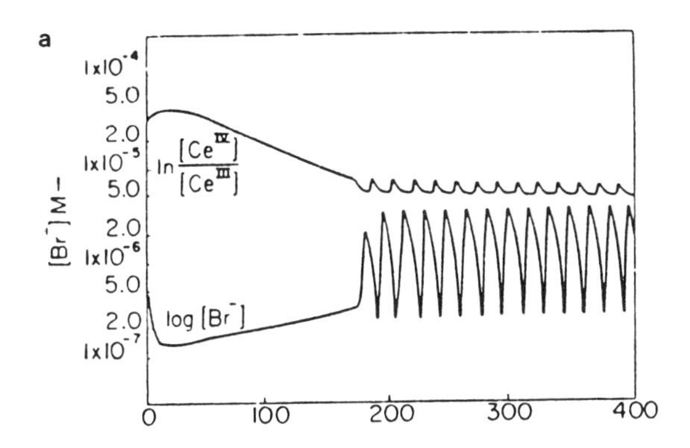

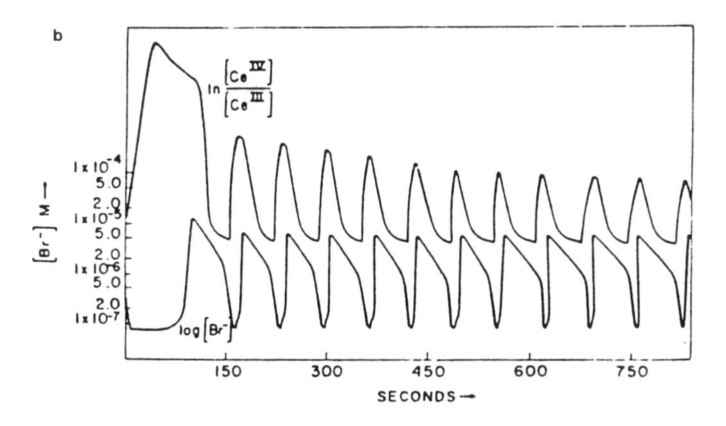

Figure 2.11. Oscillations in bromide and cerium ions in the Belousov–Zhabotinskii reaction (a) at high malonic acid content and (b) at high cerium content [from R.J. Field, E. Körös, and R.M. Noyes, *J. Am. Chem. Soc.* **94**, 8649 (1972)].

$$2HBrO_2 \rightleftarrows HOBr + BrO_3^- + H^+ \tag{B-2}$$

This process is important only when the bromide concentration is low, because a high concentration of bromide inhibits process B through reaction A-2. The autocatalytic production of $HBrO_2$ is limited by B-2. Also, the Ce^{4+} concentration is high when process B operates, in agreement with Figure 2.10, which shows that $[Ce^{4+}]$ is high when $[Br^-]$ is low.

Finally, process C recovers Ce^{3+}, as well as the Br^- ion

$$4Ce^{4+} + BrCH(COOH)_2 + 2H_2O \rightarrow Br^- + 4Ce^{3+} + HCOOH + 2CO_2 + 5H^+ \tag{C-1}$$

Through C-1, the reaction switches back to process A, which consumes the Br^- ion until the concentration is so low that process B starts to be important. This process produces the Ce^{4+} necessary for producing bromide in process C-1. Thus the system oscillates between situations of low and high Br^- content as shown in Figure 2.11.

The reaction oscillates equally well if one replaces cerium by iron, or bromine by iodine. If the redox indicator ferroin is used, the solution switches continuously between red (Fe^{2+}) and blue (Fe^{3+}), with an oscillation period between seconds and minutes, depending on the concentrations. In addition to the temporal oscillations shown in Figure 2.11, the reaction also exhibits spatial oscillations. Provided the reaction is carried out in a thin layer, there will be beautiful patterns of concentric rings travelling through the solution.

2.7.2. Oscillating Surface Reactions

Oscillating reactions between small molecules such as CO and O_2, CO and NO, and NO and H_2 on surfaces proceed through relatively simple mechanisms in which the concentration of surface vacancies usually plays an important role.

Oscillations have been observed in surface reactions also. This phenomenon in the oxidation of CO on the surfaces of palladium and platinum has been extensively studied (Conrad *et al.*, 1974; Engel and Ertl, 1978). Whether oscillations actually arise depends sensitively on the reaction temperature and the gas-phase concentrations of CO and O_2. Although different mechanistic reasons for the occurrence of oscillations exist, autocatalytic steps are always involved.

The oscillation of the CO oxidation on the (110) surface of platinum can be understood as the consequence of the structural reorganization of the surface under the influence of adsorbing molecules. As shown in Figure 2.12, the platinum (110) surface reconstructs to a less reactive structure, indicated with the surface-crystal-

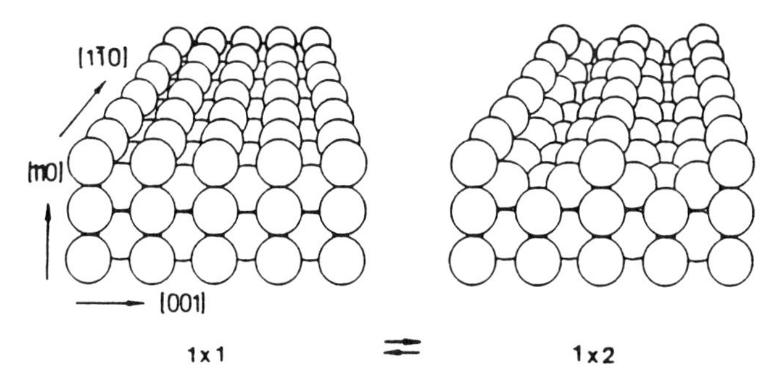

Figure 2.12. The (110) surface of platinum (left) and its (1×2) reconstruction (right).

lographic notation Pt(110)- (1×2) because the unit cell is twice as large as that of the unreconstructed surface, on which the O_2 molecule hardly adsorbs. Oscillations arise, because upon CO adsorption the surface switches to the reactive, unreconstructed (110) form on which O_2 adsorbs dissociatively and CO oxidation proceeds until all CO has been removed. The surface reconstructs to the less reactive form, and the cycle starts again. The autocatalytic feature is the reconstruction of the surface to a more reactive surface able to dissociate O_2 when CO adsorbs.

The reaction sequence of this mechanism is as follows (# and * represent sites on the unreactive and reactive surfaces, respectively). At the start, where the surface is empty and in the less reactive (1×2) form, we have the following reaction steps

$$O_2 + 2\#_{1\times2} \rightarrow 2\,O_{ads} \qquad \text{slow} \qquad \theta_O \approx 0 \tag{a}$$

$$CO + \#_{1\times2} \rightarrow CO_{ads} \qquad \theta_{CO} \text{ increases} \tag{b}$$

$$CO_{ads} + O_{ads} \rightarrow CO_2 + 2\#_{1\times2} \qquad \text{fast} \qquad r = k\theta_{CO}\,\theta_O \approx 0 \tag{c}$$

If θ_{CO} exceeds a critical value, the surface switches to the unreconstructed form, where we have the following reactions

$$O_2 + 2*_{110} \rightarrow 2\,O_{ads} \qquad \text{fast} \qquad \theta_O \text{ increases} \tag{d}$$

$$CO + *_{110} \rightarrow CO_{ads} \tag{e}$$

$$CO_{ads} + O_{ads} \rightarrow CO_2 + *_{110} \qquad \text{fast} \qquad r = k\theta_{CO}\,\theta_O > 0 \quad \theta_{CO} \text{ decreases} \tag{f}$$

If θ_{CO} decreases below the critical value, the surface reconstructs back to the unreactive form, where the oxidation reaction proceeds until all O_{ads} is removed

$$CO_{ads} + O_{ads} \rightarrow CO_2 + 2\#_{1\times2} \qquad \text{fast} \qquad \theta_O \text{ decreases to } 0 \tag{f}$$

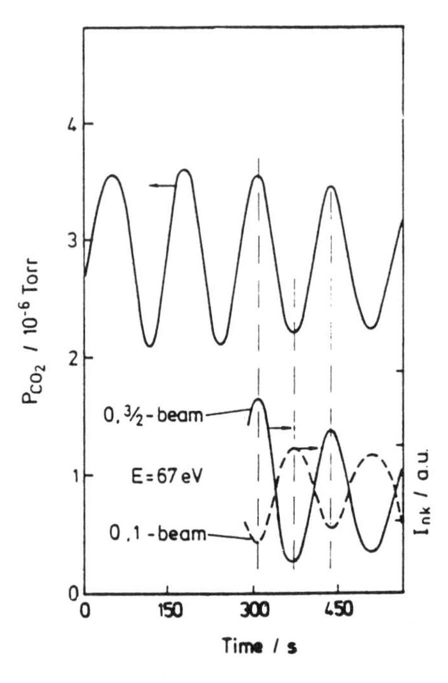

Figure 2.13. Oscillations in the $CO + O_2$ reaction over platinum single crystals. The lower curves represent the oscillations in surface structure. [from M. Eiswirth, P. Möller, K. Wetzl, R. Imbihl, and G. Ertl, *J. Chem. Phys.* **90**, 510 (1989)].

and the cycle repeats. Figure 2.13 shows the oscillations in the rate of CO_2 formation together with the intensities of two electron diffraction signals which are characteristic of the (1×1) and (1×2) reconstructions of the Pt(110) surface.

Oscillations in reactions on surfaces can also arise without structural phase transformations. The CO oxidation on Pd(110), for example, represents a case where the presence and absence of subsurface oxygen atoms explain the oscillations. The Pd(110) is highly reactive with respect to oxygen such that oxygen atoms dissolve in the outer layers. This deactivates the surface and inhibits further oxygen dissociation. However, CO can still adsorb, and it reduces the surface by removing the subsurface oxygen. The autocatalytic act of CO in this case is to remove oxygen from subsurface sites and create active sites for oxygen dissociation.

Autocatalysis can lead to explosive behavior. This has been observed in the reaction between NO and CO on the surfaces of platinum and rhodium. The overall reaction is

$$NO + CO \rightarrow \frac{1}{2}N_2 + CO_2 \tag{2.155}$$

For this reaction to proceed, empty sites have to be available so that NO can dissociate. On a completely covered surface, this is not possible and a molecule has

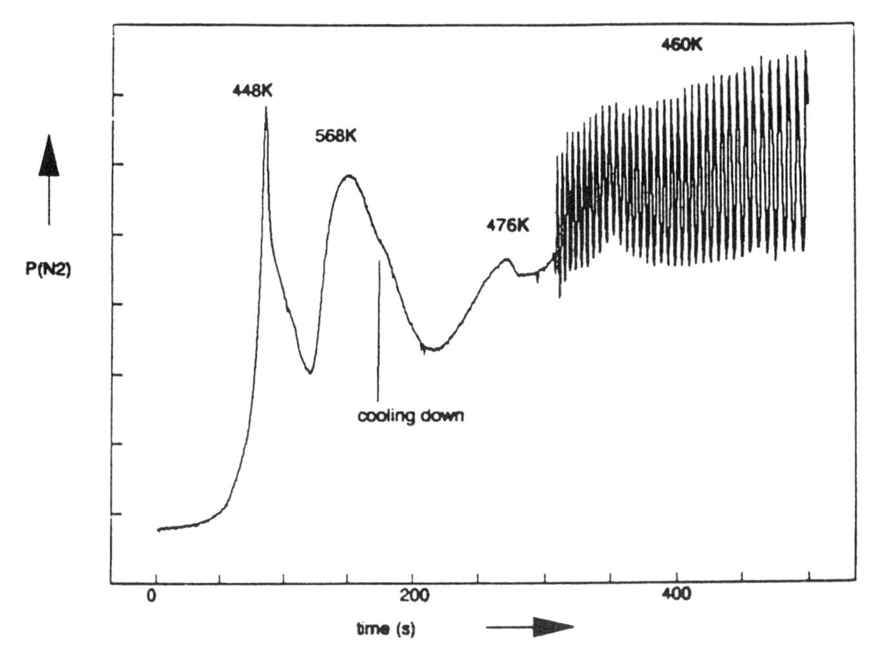

Figure 2.14. Rate of N_2 formation in the NO + H_2 reaction on the (100) surface of platinum. At low temperatures the surface is occupied by NO. During heating from room temperature to 600 K, a surface explosion takes place around 448 K. Oscillations in the production of N_2 develop on cooling down. When the temperature is kept constant at 460 K, these oscillations continue for many hours. (from J. Siera, Ph.D. thesis, University of Leiden, 1992).

to desorb to generate empty sites. Once space is available for an NO molecule to dissociate, the reaction behaves explosively

$$CO_{ads} + NO_{ads} + * \rightarrow \frac{1}{2}N_2 + CO_2 + 3* \tag{2.156}$$

The reaction is autocatalytic in the generation of free sites. Each empty site generates three new empty sites all of which initiate a new reaction event.

Figure 2.14 shows the occurrence of a surface explosion in the reaction of NO and H_2 when the temperature increases. At low temperature, the surface is mainly covered with NO. Once the temperature is high enough that desorption of NO takes place, NO may dissociate and H_2 adsorption may occur. As a result the reaction of N_{ads} to N_2 as well as the removal of O_{ads} by H_2 occur explosively. Figure 2.14 also shows that the rate of the NO + H_2 reaction starts to oscillate if the system cools down from heating to ~600 K. The oscillation period depends on the temperature at which the reacting system is held and the concentrations of NO and H_2.

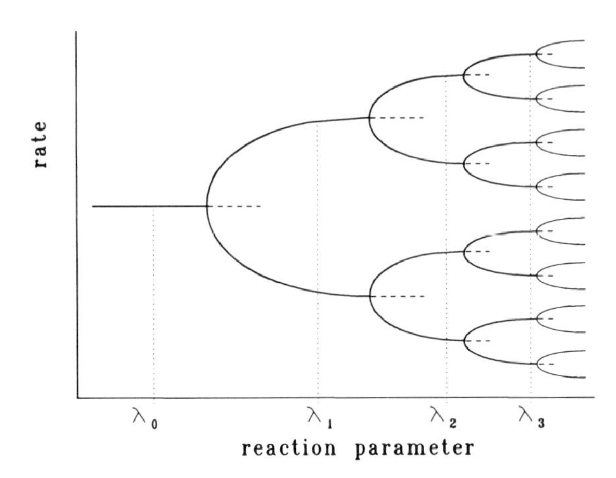

Figure 2.15. Feigenbaum diagram: if a reaction parameter (e.g. combination of concentrations and temperature) deviates sufficiently far from equilibrium conditions, the rate of the reaction may start to oscillate between several levels. At high deviations from equilibrium the system may behave chaotically.

We have only considered oscillations between two states of the reaction. Depending on the concentrations of the reactants, many more states can be realized. This is indicated in the Feigenbaum diagram of Figure 2.15. It shows the rate of a reaction as a function of a suitable reaction parameter λ (a temperature, or a ratio of concentrations). The left side of the diagram contains only one branch, corresponding to the conventional situation in kinetics. There, the reaction system is not too far from the equilibrium state, and there is only one rate of reaction for each set of reaction conditions. If the system deviates from equilibrium to the point the reaction parameter exceeds a critical value, the system becomes unstable and oscillates between two states (for example between high and low Br^- content in the Belouzov–Zhabotinsky reaction, or between the two surface reconstructions of the Pt(110) in the CO oxidation). If the reaction parameter becomes larger, more branches are available for the rate, and at λ_2 the reaction cycles between four different levels. Further to the right of the Feigenbaum diagram, there are so many levels that the system is said to exhibit chaotic behavior.

The bifurcation diagram of Figure 2.15 is often referred to as the Feigenbaum route to chaos. It applies also to reactions such as $CO + O_2$ and $NO + H_2$ on platinum, as Figure 2.16 shows. The single period oscillation between two states corresponds to point λ_1 in Figure 2.15, the double period oscillation between four states at $\lambda = \lambda_2$, a four period oscillation between eight states (λ_3) and the aperiodic, unpredictable oscillation of an essentially chaotic system.

One should not confuse chaotic with disordered. A chaotic system has all characteristics of a system that is far from equilibrium, such as a low entropy and

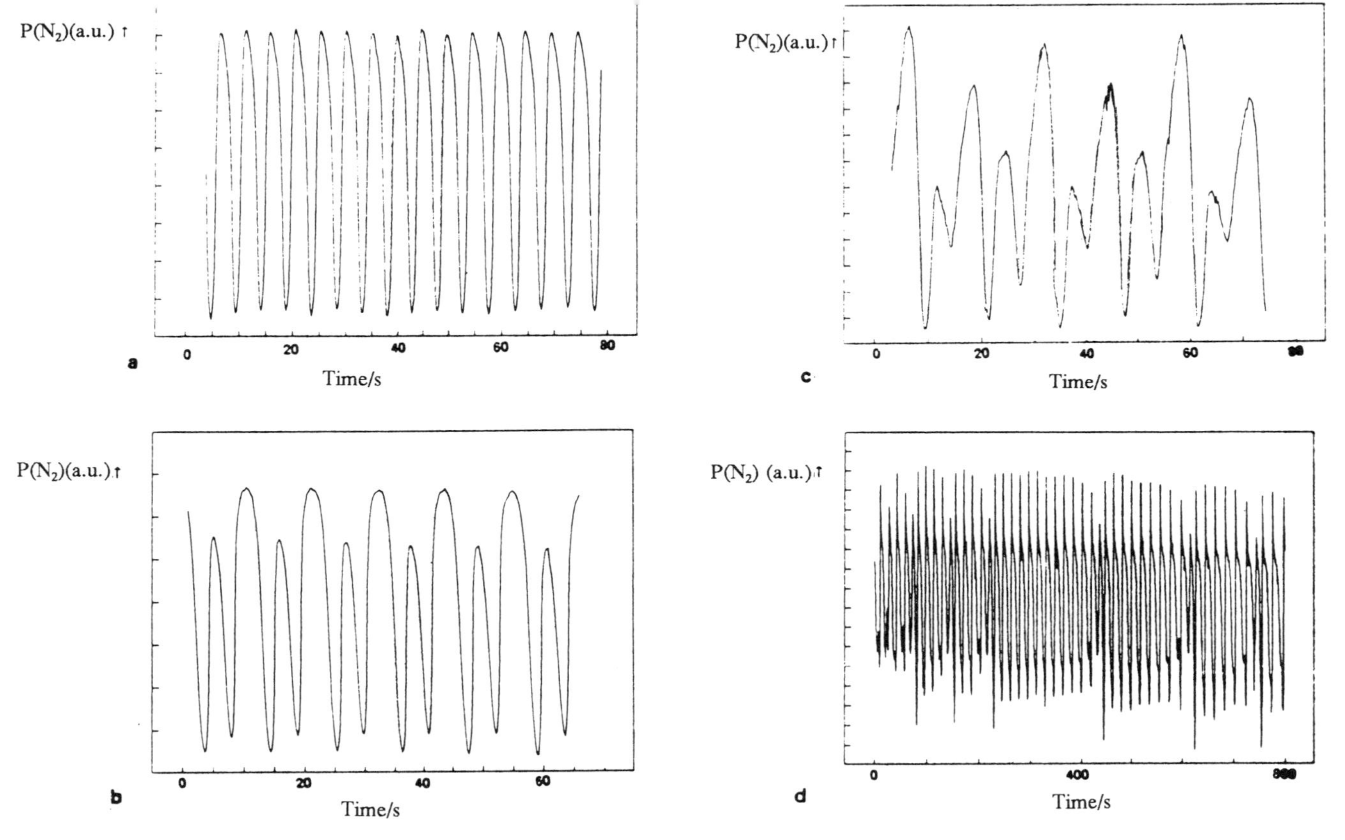

Figure 2.16. Oscillations in the rate of N_2 formation from NO and H_2 on the (100) surface of platinum, showing single, double, four-period, and chaotic oscillations [from P.D. Cobden, J. Siera, and B.E. Nieuwenhuys, *J. Vac. Sci. Technol. A*, **10**, 2487 (1992)].

a high rate of entropy production. Chaotic implies that it is not possible to predict the behavior of the system on a long term, as is the case for systems in the regime of linear thermodynamics, or even for the systems which show sustained oscillations between a small number of states, as in a chemical clock.

To summarize, chemical oscillations are very common in living organisms. Oscillations can also occur in inorganic chemistry, as in the Belouzov–Zhabotinsky reaction. However, in biochemistry oscillating systems consist of highly complex molecules which react through simple mechanisms, whereas oscillation reactions, such as the Belouzov–Zhabotinsky reaction, are based on simple molecules which react through complicated reaction schemes. In this respect, oscillating surface reactions appear to proceed through relatively simple mechanisms, implying that they form an interesting class of systems for studies of reactions under conditions that are far from equilibrium.

INTRODUCTION TO CATALYTIC REACTIONS

In many heterogeneous processes the catalytic surface provides a reaction pathway for the low temperature dissociative adsorption of reacting molecules. This step often competes with the need to desorb product molecules from the surface.

This chapter briefly discusses the mechanisms of a few important heterogeneous catalytic reactions. Also introduced are the different kinds of catalysts used and the chemisorption of a few common reactant molecules on these materials is reviewed.

Hydrogenations and oxidations form two important classes of catalytic reactions. In heterogeneous catalysis, the metals from the groups VIII and IB of the periodic system, as well as oxides or sulfides, catalyze such reactions. In view of their unique reaction mechanisms, acid-catalyzed reactions are considered as a separate class, while a fourth category is formed by reactions that are catalyzed by coordination complexes or organometallic complexes in solution, as in homogeneous catalysis. Heterogeneous catalytic reactions will be the focus, however.

Adsorption of reactants on the surface of a catalyst represents the first elementary step in a catalytic reaction cycle. Chemisorption and physisorption are two kinds of adsorption, and differ according to the type and strength of bond. In physisorption, the adsorption bond is due to the rather weak van der Waals interactions between permanent or induced dipoles. Chemisorption occurs if a real chemical bond is formed between the substrate and the adsorbate. Molecules may adsorb intact or dissociate on the surface. Catalytic reactions almost always involve the dissociation of at least one of the reacting molecules. In certain highly stable molecules, such as methane or ethane, chemisorption is not possible without the rupture of a C–H bond. Dissociation of molecules on metals leads to predominantly neutral fragments (homolytic bond splitting), whereas on oxides, the dissociation may also lead to ionic adsorbates (heterolytic splitting). Adsorption and catalysis on metals and oxides will be discussed separately.

The aim of this chapter is to give a brief overview of the chemistry involved in the most important catalytic processes, which serves as a frame of reference for the following chapters. Chapter 6 discusses a number of reactions in greater detail, to illustrate theoretical concepts that form the subject of Chapters 4 and 5.

3.1. CATALYSIS BY METALS

3.1.1. Chemisorption on Metal Surfaces

Adsorption comprises physisorption, involving van der Waals interactions, and chemisorption, in which a chemical bond forms between the adsorbate and the substrate. Chemisorption may be molecular or dissociative, depending on temperature, surface structure, and the extent to which the surface is covered by adsorbate species.

Metals of catalytic interest are found in Groups VIII and I-B of the Periodic Table (Table 3.1). These metals are stable against oxidation or carburization under the conditions prevailing in most catalytic reactions. Figure 3.1 summarizes the structure of the most stable surface planes of the metals. In general, adsorption heats of atoms increase in the Periodic Table starting from the left and up. The probability that a molecule dissociates follows the same trend. Metals to the left of iron become increasingly more reactive and form very strong adsorption bonds. Consequently, reactions between adsorbed species are largely prohibited.

Adsorption bonds are highly localized and occur between the adsorbate and the nearest neighbors in the surface. The way molecules chemisorb on metal surfaces depends on a number of parameters such as the reactivity of the metal, the temperature at which adsorption takes place, the structure of the substrate surface and the surface coverage of the adsorbate.

TABLE 3.1. The Catalytically Active Metals and their Lattice Structure

Group VIII			I-B
Fe	Co	Ni	Cu
bcc	*hcp*	*fcc*	*fcc*
Ru	Rh	Pd	Ag
hcp	*fcc*	*fcc*	*fcc*
Os	Ir	Pt	Au
hcp	*fcc*	*fcc*	*fcc*

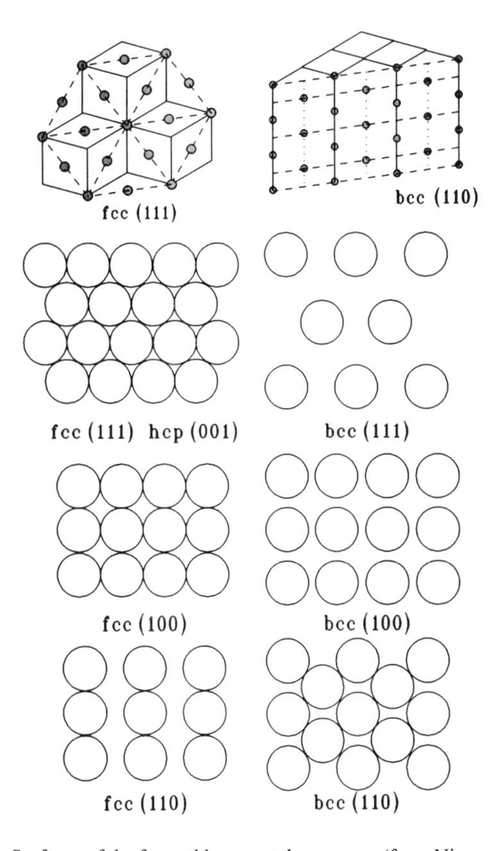

Figure 3.1. Surfaces of the fcc and bcc crystal structures (from Niemantsverdriet, 1993).

Temperature: Almost all molecules (with the possible exception of hydrogen) adsorb molecularly at sufficiently low temperature. When the temperature is raised, the molecule may either desorb or dissociate. In the latter case, the adatoms may recombine at higher temperatures and desorb in a second-order reaction.

Structure: Smooth, densely packed metal surfaces, such as the fcc (111) and (100) and the bcc (110) structures shown in Figure 3.1 are relatively unreactive, while more open surfaces, such as fcc (110) and bcc (111), are much more reactive. These surfaces form stronger bonds with adsorbates, and the probability for molecules to dissociate is significantly higher. Also the presence of steps, kinks, or defects in general, considerably increases the reactivity of a surface. The adsorption heat of almost any molecule therefore depends on the crystal plane on which it adsorbs. Because a catalyst consists of particles which expose many different surfaces and all kinds of defects, the heat of adsorption on a catalyst represents an ill-defined, averaged value.

TABLE 3.2. Dissociation Energies of Bonds in Gas Phase Molecules

Molecule	Dissociation energy (kJ/mole)	Molecule	Bond	Dissociation energy (kJ/mole)
H_2	436	CH_4	C–H	435
O_2	497	C_2H_6	C–H	410
NO	626		C–C	368
CO	1074	C_2H_4	C C	724
N_2	949	C_2H_2	C–C	962
		CH_3OH	C–H	393
			O–H	435
		H_2O	O–H	498

Coverage: All the information we present in this chapter is valid for adsorption on otherwise empty surfaces. As discussed extensively in Chapters 5 and 6, interactions between adsorbed species may have a large effect on the heat of adsorption, in particular when the adsorbate acquires a net positive or negative charge, or when it possesses a dipole moment. In addition, the mere fact that adjacent adsorption sites are occupied may prevent a molecule to dissociate. Lateral interactions also determine the ordering of adsorbates in periodic structures or the segregation of alike species in two-dimensional islands.

Reviewed here are some of the most important adsorption characteristics of these molecules. Dissociation energies are listed in Table 3.2. On the majority of the catalytically active metals the heats of adsorption increase in the following sequence:

$$N_2 < CO_2 < H_2 < CO \approx NO < O_2 \tag{3.1}$$

Hydrogen dissociates on all metals already at low temperatures; only on copper is the dissociation activated. The metal-hydrogen bond strength varies little with the substrate metal, differences in surface structure are more important. As the recombinative desorption of hydrogen atoms is a rapid process, the surface coverage of hydrogen is usually low at typical reaction temperatures.

Oxygen dissociates on all Group VIII/Ib metals, although on Pt, Ag, and Au, oxygen can also exist in the molecularly bound form. Oxygen atoms almost always bind in sites of high symmetry with threefold or fourfold coordination to the substrate metal. In some metals, O atoms can also be accommodated in deeper layers of the metal. The existence of subsurface oxygen in silver, for example, is well documented.

Nitrogen interacts weakly with most of the Group VIII/Ib metals. Dissociative adsorption is only possible on iron, ruthenium, and on metals toward the left of the Periodic Table. This behavior correlates well with the enthalpy of metal nitride formation which is positive for copper, nickel, and cobalt nitrides, but negative for iron and its more reactive neighbors to the left. Promoter atoms, such as potassium or cesium, are used to increase the dissociation probability of N_2.

Carbon monoxide adsorbs molecularly at low temperatures on all the catalytically relevant metals. When the temperature increases, it dissociates on iron around room temperature, and on cobalt, nickel, and ruthenium at higher temperatures. Rhodium represents a borderline case; reaction conditions exist where molecular CO coexists with carbon and oxygen atoms from dissociated CO. When the temperature is sufficiently high, CO can react with adsorbed oxygen to form CO_2, which desorbs instantaneously while carbon remains on the surface. This is called the Boudouard reaction.

Molecular CO adsorbs with the carbon atom toward the metal and the C-O bond perpendicular to the surface and coordinates in several geometries (identified by the CO stretch frequencies in vibrational spectroscopy). Linear CO is adsorbed on top of a metal atom, bridged between two- and threefold CO between three metal atoms of the substrate. On the dense (111) and (100) surfaces of the fcc metals, CO prefers the linear and bridged modes over the high coordination of the threefold adsorption geometry. On rhodium and platinum, CO adsorbs more strongly in the linear mode than in the bridged mode. On nickel and palladium, the bridged mode forms the stronger bond.

Nitrogen oxide (NO) is significantly more reactive than CO due to its electron in the antibonding $2\pi^*$ orbital, which is empty in gas-phase CO. As a result, NO dissociates on more metals than CO does. Although a complete picture of CO is not yet available, it is certain that NO, provided the coverage is low, readily dissociates on all surfaces of rhodium and nickel at room temperature or lower, and also on the more corrugated surfaces of platinum. Nitrogen atoms desorb by recombination into N_2, while recombination with oxygen adatoms has not been observed.

Typical product molecules in many catalytic reactions such as CO_2 and H_2O interact weakly with most metal surfaces and desorb readily upon formation under reaction conditions. Ammonia, on the other hand, decomposes on almost all metals.

Hydrocarbon interaction with metal surfaces almost always involves the breaking of C–H or sometimes C–C bonds, and can lead to a multitude of hydrocarbon fragment species, depending on temperature, metal, and structure of the surface. The fully saturated alkanes interact only weakly with metal surfaces. At low temperatures (below 375 K on smooth surfaces) reaction is limited to the exchange of hydrogens in the alkane with hydrogen or deuterium atoms on the surface. Formation of a stronger chemisorption bond takes place if at least one C–H bond breaks, to form an alkyl and a hydrogen atom on the surface, which is only possible at higher temperatures on the densely packed metal surfaces (Ni: approximately 375 K and higher). At still higher temperatures (Ni: around 440 K), C–C bond breaking and further dehydrogenation become likely.

In alkanes, the C–C bonds are weaker than the C–H bonds (Table 3.2). Nonetheless on a metal surface the C–H bonds dissociate first. The reason is a steric one. Once hydrogen atoms have been removed, the molecular carbon atoms and surface metal atoms approach so close that the C–C bond becomes activated. Steps, kinks, or other defects increase the reactivity of the surface toward alkane activation. Also the partial pressure of hydrogen above the surface, which controls the coverage of H atoms on the surface, plays an important role in the composition of the adsorbed hydrocarbon fragments.

Alkenes, alkynes, and aromatic hydrocarbons are much more reactive than the alkanes. Their adsorption requires hardly any activation and the molecules easily establish chemisorption bonds with a surface under vacuum conditions, and at low temperatures. Aromatic hydrocarbons, such as benzene and toluene, usually adsorb with the ring parallel to the surface, but alkenes and alkynes show more variety with respect to surface bonding.

Ethylene, when adsorbed on platinum at low temperatures (<200 K), binds to two adjacent Pt atoms with a single intramolecular C–C bond (di-σ bonding). While on metals such as Cu, Pd, and Ni, the ethylene binds via an interaction between the double C=C bond and a single metal atom (π-bonding). Just below room temperature, the di-σ-coordinated C_2H_4 on platinum splits one H-atom off and rearranges to the strongly bound ethylidyne, C–CH_3, in which the bare carbon atom binds to three Pt atoms in the surface. Upon further heating to ~ 400 K, the C–C bond breaks and the surface becomes covered by CH_x species. At higher temperatures, these fragments dehydrogenate further and polymerize into the largely unreactive, graphitic carbon structures, which contribute to the deactivation of platinum catalysts in industrial processes. The process is readily followed by means of thermal desorption spectroscopy (Figure 3.2).

Different ethylene-derived species, such as vinyl or acetylene, have been observed on other metals. On the (100) surface of iron, for example, the di-σ-bound C_2H_4 dehydrogenates to acetylene around 200 K, while at about 400 K the triple C≡C bond breaks to yield CH and CH_2 fragments, which eventually release their hydrogen to form carbidic or even graphitic carbon at elevated temperatures.

Surface irregularities play a crucial role in the decomposition of hydrocarbons. While acetylene (C_2H_2) remains intact on the smooth, densely packed (111) surface of nickel and shows no tendency to form CH fragments below 400 K, the molecule dehydrogenates completely on a stepped nickel surface at 150 K, while the C=C bond breaks already at 180 K.

Chemisorption of alcohols on the transition metals involves rupture of the O-H bond and the subsequent formation of an alkoxy species and a hydrogen atom. Methanol, for instance, readily forms the methoxy group (OCH_3). Depending on the reactivity of the substrate metal, the methoxy species decomposes further to adsorb formaldehyde (CHOH), formyl species (COH), and eventually CO.

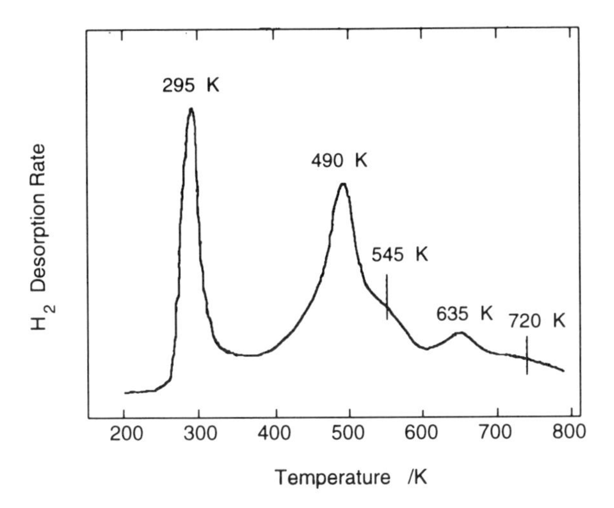

Figure 3.2. Temperature programmed desorption spectra of hydrogen illustrate the decomposition of alkenes adsorbed on platinum (111) at 120 K (adapted from Creighton and White, 1983).

The sequential decomposition is a general characteristic of the interaction between hydrocarbons and metals. Although the total decomposition of ethylene on a metal as symbolized by the reaction

$$C_2H_4 + 6\ M \rightarrow 2\ M\text{–}C + 4\ M\text{–}H \tag{3.2}$$

has a negative change in free energy and is thus thermodynamically allowed, the decomposition is sequential. Each rupture of a bond requires its own characteristic activation energy, causing the existence of characteristic C_xH_y species in well-separated temperature regimes.

3.1.2. Hydrogenation and Related Reactions

3.1.2.1. Ammonia Synthesis

Ammonia is an important intermediate in the production of fertilizer and explosives. As explained in Chapter 1, the development of the ammonia synthesis represented a formidable technological achievement and a significant stimulus for the development of catalysis as a science. A plant for the synthesis of ammonia contains several catalytic processes. It starts with the production of hydrogen, by the nickel-catalyzed steam-reforming of natural gas

$$CH_4 + H_2O \rightarrow CO + 3\ H_2 \tag{3.3}$$

followed by the water–gas shift reaction, in which CO is used to reduce more steam

$$CO + H_2O \rightarrow CO_2 + H_2 \tag{3.4}$$

The latter is carried out in two steps: 1) with a cheap iron oxide catalyst and 2) with a more expensive supported copper-zinc oxide catalyst. After removal of the CO_2 by pressure washing, residual traces of CO are hydrogenated in the methanation reaction, the reverse of the steam reforming (3.3), on a supported nickel catalyst. Nitrogen is obtained from the air.

The actual ammonia synthesis reaction is an equilibrium

$$N_2 + 3\,H_2 \leftrightarrows 2\,NH_3 + \text{heat} \tag{3.5}$$

As high pressures and low temperatures shift the equilibrium to the right, the process calls for an active catalyst capable of driving the reaction to equilibrium at relatively low temperatures.

Iron forms sufficiently strong bonds with nitrogen atoms to overcome the high bond energy of the N_2 molecule. Ammonia is the end product of a chain of surface association reactions of adsorbed nitrogen and hydrogen atoms that proceeds via the formation of adsorbed NH_x intermediates. This is expressed by the following simplified scheme of reactions

$$\begin{aligned} N_2 + 2* &\leftrightarrows 2N_{ads} \quad \text{(r.d.s.)} \\ H_2 + 2* &\leftrightarrows 2H_{ads} \\ N_{ads} + 3H_{ads} &\leftrightarrows NH_{3,ads} + 3* \\ NH_{3,ads} &\leftrightarrows NH_3 + * \end{aligned} \tag{3.6}$$

These reactions lead to a rate of reaction of the form

$$V\frac{d[NH_3]}{dt} = N_s \frac{k_1^f[N_2][H_2]^3 - k_1^b K_\alpha^2 [NH_3]^2}{([H_2]^{3/2} + (K_4^{-1}[H_2]^{3/2} + K_\alpha)[NH_3])^2} \tag{3.7}$$

in which $K_\alpha = 1/(K_3\,K_2^{3/2}K_4)$ and $(K_2\,[H_2])^{1/2}$ is assumed to be small.

The dissociation of N_2 is a slow reaction step. It is suppressed by competitive adsorption of NH_3 that blocks sites needed for nitrogen dissociation. Promoters, such as potassium, are added to enhance the rate of nitrogen dissociation and to weaken the interaction of NH_3 with the catalyst surface. Reduction of the surface coverage of adsorbed NH_3 enhances the overall rate of reaction, as indicated by the negative order in ammonia in expression (3.7). Details of the reaction mechanism are given in Chapter 6, Section 6.6.3.

3.1.2.2. Synthesis Gas Conversion

The term synthesis gas almost always refers to mixtures of CO and H_2 in various proportions. Synthesis gas is obtained from hydrocarbons, either by the nickel-catalyzed, steam-reforming reaction (3.3), partial combustion with oxygen and reaction with steam, or by the gasification of coal (Figure 3.3).

The catalytic reaction between CO and H_2 to hydrocarbons and methanol is mechanistically related to the ammonia synthesis. Many different products, ranging from methane to the longer hydrocarbons encountered in diesel oil or in waxes, as well as oxygenated products such as methanol, ethanol, and acetaldehyde can be formed depending on the choice of the catalyst and the reaction conditions.

In order to produce methanol the catalyst should only dissociate the hydrogen but leave the carbon monoxide intact. Metals such as copper (in practice promoted with ZnO) and palladium as well as several alloys based on noble group VIII metals fulfill these requirements. Iron, cobalt, nickel, and ruthenium, on the other hand, are active for the production of hydrocarbons, because in contrast to copper, these metals easily dissociate CO. Nickel is a selective catalyst for methane formation. Carbidic carbon formed on the surface of the catalyst is hydrogenated to methane. The oxygen atoms from dissociated CO react with CO to CO_2 or with H-atoms to water. The conversion of CO and H_2 to higher hydrocarbons (on Fe, Co, and Ru) is called the Fischer–Tropsch reaction. The Fischer–Tropsch process provides a way to produce liquid fuels from coal or natural gas.

The following scheme accounts for the methanation reaction

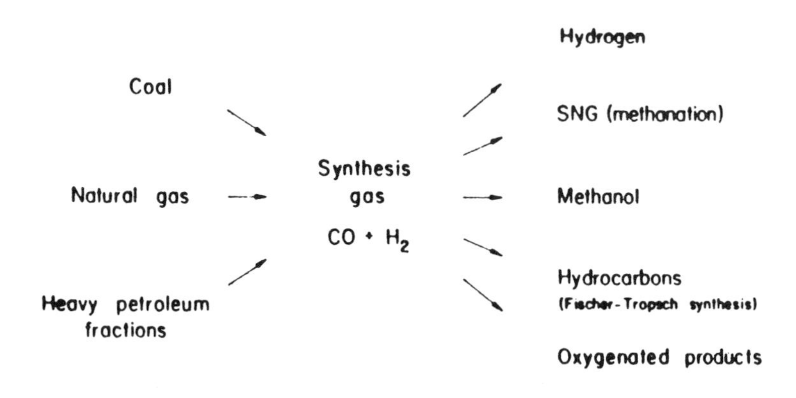

Figure 3.3. Production and use of synthesis gas, CO + H_2.

$$
\begin{aligned}
&CO + * \leftrightarrows CO_{ads} && (a)\\
&CO_{ads} + * \leftrightarrows C_{ads} + O_{ads} && (b)\\
&H_2 + 2* \leftrightarrows 2H_{ads} && (c)\\
&C_{ads}\text{+Hads} \leftrightarrows CH_{ads} + * && (d)\\
&CH_{ads} + H_{ads} \leftrightarrows CH_{2,ads} + * && (e)\\
&CH_{2,ads} + H_{ads} \leftrightarrows CH_{3,ads} + * && (f)\\
&CH_{3,ads} + H_{ads} \leftrightarrows CH_4 + 2* && (g)\\
&O_{ads} + 2H_{ads} \leftrightarrows H_2O + 3* && (h)
\end{aligned}
\tag{3.8}
$$

CO dissociation is an activated reaction step, so the reactive surface is covered with CO. Whereas adsorption of CO requires one surface site, dissociation of CO requires empty surface sites (6.5). Upon dissociation of a molecule, two adatoms are generated that now will occupy two surface sites. When a surface is completely covered by CO, at least one CO molecule beside the reacting CO has to be removed in order to make dissociation of the other CO molecules possible. The overall effect is that the CO hydrogenation reactions have a negative reaction order in CO. The order in hydrogen is positive, however. Many promoters added to Fischer–Tropsch catalysts enhance the rate by an increase of the rate of CO dissociation (Section 6.6.5).

In spite of the factors that have a negative effect on the probability that the CO bond breaks, CO dissociation is not thc ratc-determining step. If this were the case, then the rate would not exhibit a positive order in hydrogen. If the reaction between an adsorbed carbon and the first hydrogen atom were rate determining, and all subsequent reactions fast, it would be

$$
\theta_{CO} = K_1[CO]\theta_* \; ; \; \theta_C = \sqrt{K_1 K_2[CO]}\, \theta_* \; ; \; \theta_H = \sqrt{K_3[H_2]}\, \theta_* \tag{3.9}
$$

and the rate would become proportional to the coverages of C and H

$$
r = V\frac{d[CH_4]}{dt} = \frac{N\, k_4\, (K_1K_2K_3)^{1/2}\, [CO]^{1/2}[H_2]^{1/2}}{(1 + K_1[CO] + 2(K_1K_2[CO])^{1/2} + (K_3[H_2])^{1/2})^2} \tag{3.10}
$$

In this model, the order in CO falls between –3/2 and 1/2, while the order in hydrogen would be 1/2 at most, corresponding to the limit of weak adsorption.

Taking the reaction between CH_{ads} and H_{ads} as rate limiting, then step (4) yields the coverage of CH_{ads}

$$\theta_{CH} = (K_1K_2K_3)^{1/2}K_4\,[CO]^{1/2}[H_2]^{1/2}\theta_* \tag{3.11}$$

and the rate equals

$$r = \frac{N\,k_5\,(K_1K_2)^{1/2}\,K_3K_4\,[CO]^{1/2}\,[H_2]}{(1 + K_1[CO] + 2(K_1K_2[CO])^{1/2} + (K_3[H_2])^{1/2} + (K_1K_2K_3)^{1/2}K_4[CO]^{1/2}[H_2]^{1/2})^2} \tag{3.12}$$

This expression can indeed account for a positive, first order in hydrogen and a negative or close to zero order in CO as is experimentally observed. The expression is also valid for the Fischer–Tropsch synthesis of higher hydrocarbons. In this case the scheme of (3.8) has to be extended with chain-growth reactions, as discussed in Section 6.6.5. How to control the selectivity of this process is a key issue in CO hydrogenation catalysis. Methane and methanol are the only products that can be obtained with 100% selectivity.

3.1.2.3. Hydrocarbon Activation

Nickel, palladium, and platinum are metals commonly used as catalysts in reactions of hydrocarbons. They provide an optimum interaction with adsorbed hydrocarbon fragments. As mentioned in Chapter 1, the hydrogenation of unsaturated hydrocarbons over nickel catalysts was the first industrial application of catalytic hydrogenation. The process, known as fat hardening, results in an increase of the boiling point of the fatty oils and is used on a large scale in the production of margarine. Nickel particles dispersed on an inert SiO_2 support are used as the catalyst.

The heat of adsorption of an alkene is larger than the heat of dissociative adsorption of hydrogen. Therefore, similar to the previously mentioned CO hydrogenation, the reaction rate has a negative order in the alkene at low temperature. A positive order is observed at high temperatures, when the equilibrium constant for the adsorption of the alkene is low enough to also lower the surface concentration. During reaction the metal catalyst surface becomes covered with a carbonaceous surface layer. The simplest unsaturated hydrocarbon is ethylene. The kinetics of its catalytic hydrogenation has been discussed in detail in Chapter 2.

Hydrogenations are usually carried out at low temperatures because the reaction is exothermic. Low temperature hydrogenations are also of great importance in the production of fine chemicals, where the reaction is often performed in the liquid phase, with carbon-supported platinum or palladium or with organometallic complexes as the catalyst. Materials that catalyze hydrogenation are also active in dehydrogenation, which becomes thermodynamically favored at higher tempera-

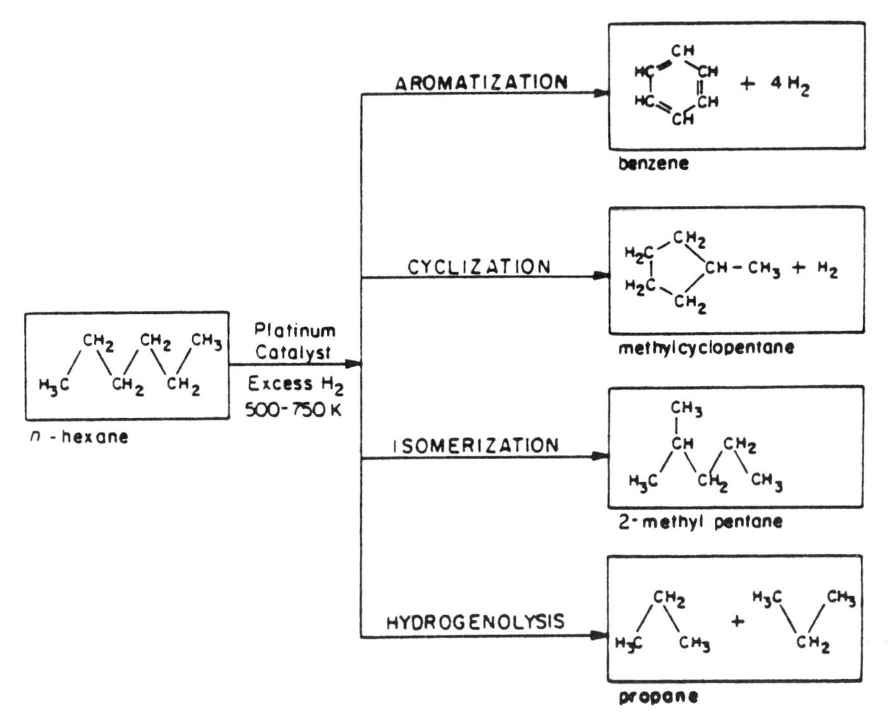

Figure 3.4. Reforming reactions of n-hexane on platinum; n-hexane is the smallest hydrocarbon which can undergo all these reactions (from Somorjai, 1981).

tures. However, metals are by no means unique in their ability to dehydrogenate, oxides are in use also.

Platinum is probably the most versatile catalyst for hydrocarbon reactions. It can activate hydrogenation, dehydrogenation, hydrogenolysis, isomerization, and aromatization. Catalytic reforming represents a large-scale process in the oil industry where all these properties of platinum are exploited. The aim of the reaction is to improve the octane number of paraffin mixtures, in order to make these suitable for use as gasoline. The smallest molecule that can undergo all of these reactions is n-hexane (Figure 3.4). Briefly described are a few examples for which the kinetics are well established.

In the hydrogenolysis reaction, rupture of the C–C bonds occurs and consecutive hydrogenation of adsorbed molecule fragments gives hydrocarbons containing less carbon atoms than the reactant molecule. The greater the metal surface reactivity, the more significant the hydrogenolysis reaction becomes. Platinum is the preferred reforming catalyst since its hydrogenolysis activity is least of those transition metals that are able to activate C–H and C–C bonds. At low temperatures the hydrogenolysis reaction has an order negative in hydrogen pressure resulting from competitive adsorption of alkane molecules with hydrogen.

The simplest hydrogenolysis is that of ethane. It is catalyzed by almost all Group VIII metals. As explained before, steric hindrance is the reason that the C–C bond, although weaker than a C–H bond, cannot be activated unless the stronger C–H bond is activated first. Thus the following mechanism, valid for ethane hydrogenolysis on nickel is considered

$$
\begin{aligned}
& C_2H_6 + 2* \leftrightarrows C_2H_{5,ads} + H_{ads} && (a) \\
& H_2 + 2* \leftrightarrows 2H_{ads} && (b) \\
& C_2H_{5,ads} + H_{ads} \leftrightarrows C_2H_{2,ads} + 2H_2 + * && (c) \\
& C_2H_{2,ads} + * \rightarrow 2CH_{ads} && (d,\ rds) \\
& CH_{ads} + 3H_{ads} \rightarrow CH_4 + 4* && (e,\ fast)
\end{aligned}
\tag{3.13}
$$

The equilibrium reactions (a)–(c) enable the determination of the relevant surface coverages

$$
\begin{aligned}
\theta_H &= K_2^{1/2}[H_2]^{1/2}\theta_* \\
\theta_{C_2H_2} &= K_1K_3[C_2H_6][H_2]^{-2}\theta_* \\
\theta_{C_2H_5} &= K_1K_2^{-1/2}[C_2H_6][H_2]^{-1/2}\theta_*
\end{aligned}
\tag{3.14}
$$

and the coverage of free sites follows from the condition

$$\theta_* + \theta_H + \theta_{C_2H_2} + \theta_{C_2H_5} = 1 \tag{3.15}$$

This leads to the following expression for the rate

$$
\begin{aligned}
r &= V\frac{d[CH_4]}{dt} \\
&= \frac{2Nk_4K_1K_3[C_2H_6][H_2]^{-2}}{(1 + (K_2[H_2])^{1/2} + K_1K_3[C_2H_6][H_2]^{-2} + K_1K_2^{-1/2}[C_2H_6][H_2]^{-1/2})^2}
\end{aligned}
\tag{3.16}
$$

The expression agrees with experimental data, which have yielded an order of one in ethane and –2.4 in hydrogen for nickel at 450 K. It appears that the reaction

TABLE 3.3. Kinetic Parameters of the Ethane Hydrogenolysis Reaction in Terms of the Power Rate Law $r = k\,[C_2H_6]^n[H_2]^m$ [a,b]

Catalyst	T (*K*)	*x*	*a*	*n*	*m*	$-na$
Co	492	4	1	1.0	–0.8	–1.0
Ni	450	2	2	1.0	–2.4	–2.0
Ru	461	2	2	0.8	–1.3	–1.6
Rh	487	0	3	0.8	–2.2	–2.4
Pd	627	0	3	0.9	–2.5	–2.7
Os	425	2	2	0.6	–1.2	–1.2
Ir	483	2	2	0.7	–1.6	–1.4
Pt	630	0	3	0.9	–2.5	–2.7

[a]From Sinfelt, 1981.
[b]See also 3.17 and 3.18 for the meaning of x and $-na$.

mechanism is different on almost each Group VIII metal catalyst, the essential difference being the extent of dehydrogenation occurring in step (3) of the mechanism. In general this reaction can be written as

$$C_2H_{5,ads} + H_{ads} \leftrightarrows C_2H_{x,ads} + aH_2 + * \quad a = 3 - \frac{x}{2} \tag{3.17}$$

which, in the limit of weak adsorption, leads to the rate expression

$$r \approx k[C_2H_6]^n[H_2]^{-na} \tag{3.18}$$

The data in Table 3.3 show that the Group VIII metals differ significantly in the extent to which ethane is dehydrogenated before the C–C bond breaks.

An example involving a larger hydrocarbon molecule is provided by the dehydrogenation (aromatization) of methyl cyclohexane ($C_6H_{11}CH_3$) to toluene ($C_6H_5CH_3$) catalyzed by platinum. Methyl cyclohexane will be abbreviated as MCH. According to Sinfelt (1981), the mechanism can be written as a sequence of reactions in the forward direction only

$$\begin{aligned} &MCH + * \rightarrow MCH_{ads} && (a) \\ &MCH_{ads} \rightarrow Toluene_{ads} + 3H_2 && (b) \\ &Toluene_{ads} \rightarrow Toluene + * && (c) \end{aligned} \tag{3.19}$$

If the reaction runs under steady-state conditions, with adsorbed toluene as the majority species and desorption of toluene as the rate-determining step, the rate of reaction becomes

$$r = V\frac{d[\mathrm{T}]}{dt} = \mathrm{N}k_3\theta_\mathrm{T} = \frac{\mathrm{N}k_1[\mathrm{MCH}]}{1 + (k_1/k_3)[\mathrm{MCH}]} \tag{3.20}$$

where T stands for toluene.

Isomerization reactions are also catalyzed by platinum, but acid sites of alumina or zeolitic supports have higher activity for this reaction, as covered later in this chapter.

3.1.2.4. Oxidation

The selective oxidation of NH_3 to NO is catalyzed by Pt/Rh. Called Ostwald reaction, it runs at high temperatures and is an important step in the conversion of nitrogen to nitrates via ammonia. Another catalytic oxidation process in inorganic chemistry is the oxidation of SO_2 to SO_3 to produce sulfuric acid. Platinum or vanadium oxide are the preferred catalysts. The catalyst has to be able to dissociate oxygen, and bulk sulfate formation, deactivating the catalyst, has to be suppressed.

On silver the dependency of the nature of adsorbed oxygen atoms on surface coverage is exploited for selective ethylene epoxidation. Ethylene epoxide is an intermediate for the production of glycol, used as antifreeze. Oxygen atoms covering silver at a low concentration are nucleophilic. These oxygen atoms prefer activation of CH over insertion into the double bond of ethylene. In this state the main reaction with ethylene is total combustion.

Only silver covered with a high concentration of oxygen, with a surface stoichiometry close to that of AgO, is a selective catalyst for the epoxidation reaction (Figure 3.5). High surface concentration enables adsorbed oxygen to become electrophilic and insert into the ethylene π-bond. Unlike ethylene oxide, the epoxide of propylene cannot be obtained by direct catalytic oxidation. The chlorohydrin route is still in use, although it is gradually being replaced by indirect oxidation routes using organic hydroperoxides.

The nucleophilic nature of oxygen atoms, when adsorbed to the metallic silver surface, is exploited for methanol oxidation. At a high temperature silver is an efficient catalyst for the conversion of methanol to formaldehyde.

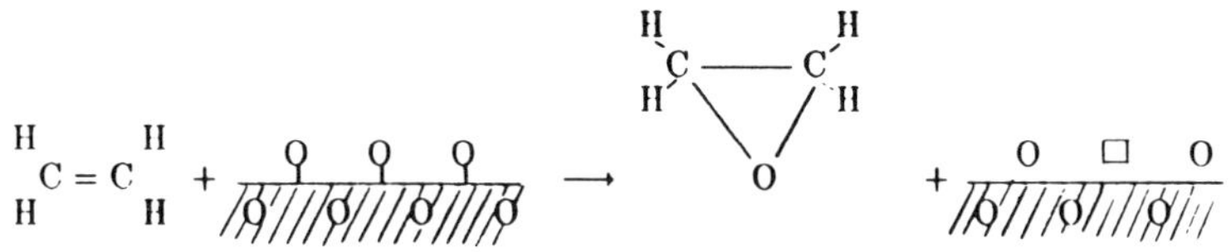

Figure 3.5. The epoxidation of ethylene on silver.

Platinum and palladium have high activities for total oxidation. This property is exploited in automotive exhaust catalysis. Automobile exhaust contains toxic gases such as CO, NO, and hydrocarbons which contribute to formation of photochemical smog and acid rain. Since 1978, catalysts based on platinum, rhodium, and sometimes palladium, supported on a monolithic carrier, are applied to convert exhaust gases to less harmful products. The so-called threeway catalyst enables the following three overall reactions

$$
\begin{aligned}
&CO + \tfrac{1}{2}O_2 \rightarrow CO_2 \\
&CO + NO \rightarrow CO_2 + \tfrac{1}{2}N_2 \\
&C_xH_y + O_2 \rightarrow CO_2 + H_2O
\end{aligned}
\tag{3.21}
$$

Other reactions occur as well: Cracking and dehydrogenation of the hydrocarbons produces hydrogen, which assists in removing oxygen from the catalyst.

Platinum is mainly responsible for the oxidation reactions as the metal has a favorable activity for these reactions at low temperatures. The kinetics of the CO oxidation is discussed in Section 6.1. The reaction between NO and CO is catalyzed by rhodium, which has excellent activity for the dissociation of NO. Rate-limiting

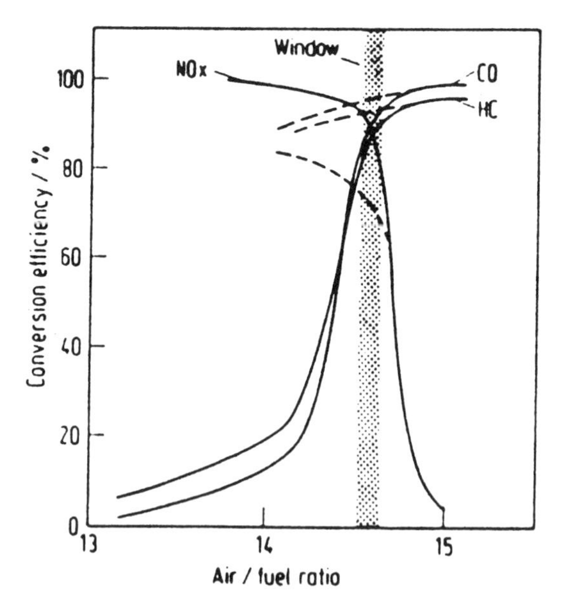

Figure 3.6. The simultaneous conversion of CO, NO, and hydrocarbons by a three way catalyst is only possible in a narrow window of air/fuel ratios (from Taylor, 1987).

is the removal of oxygen or nitrogen from the surface. The dissociation reaction is suppressed by the presence of excess oxygen.

In order to let the three reactions proceed simultaneously, the composition of the exhaust gas needs to be properly adjusted to an air-to-fuel ratio of 14.7 (Figure 3.6). At higher oxygen content, the CO oxidation reaction consumes too much CO and hence NO conversion fails. If, however, the oxygen content is too low, all NO will be converted, but the removal hydrocarbons and CO will not be complete. Oxygen sensors are used to maintain the proper balance of fuel and air.

3.2. CATALYSIS BY OXIDES

3.2.1. Chemisorption on Oxides

Characteristic features of catalysis by oxides are the involvement of acid-base properties of adsorbate and substrate, causing the heterolytic dissociation of molecules that would yield neutral species on metals, and the active participation of lattice oxygen as a reactant in oxidation reactions. The reactivity of oxide surfaces is often determined by point defects and the extent to which cations are exposed to the gas phase.

The adsorption of gases on the surfaces of oxides differs greatly from that on metal surfaces. As oxides are ionic substances, interactions involve the acid–base properties of substrate and adsorbate, while interactions between uncompensated charges at the surface and dipole moments of the adsorbing molecules also play a role. In order to create a surface of an oxide MO_x with M^{n+} cations and O^{2-} anions, M–O bonds have to be broken, with the result that the surface contains uncompensated charges. The surface cations possess excess positive charge and thus behave as Lewis acids (electron acceptors) while the oxygen atoms have excess negative charge and behave as Brønsted bases.[†] As a result of the ionic nature of the oxide surface, many molecules undergo heterolytic splitting, whereas they would dissociate in neutral radicals on the surface of a metal. Abstraction of a proton from a hydrogen-containing molecule to form a surface hydroxyl is a very general phenomenon in oxide surface chemistry.

[†]In Brønsted's concept of basicity, every species that can accept a proton (i.e. NH_3, OH^-, or a surface oxygen with uncompensated negative charge) is called a base. A Brønsted acid is a proton donor. Lewis' concept of acids and bases considers an entity capable of accepting electron density (such as surface cations with excess positive charge) as an acid, while electron donors, i.e. molecules which have lone pairs of electrons, such as NH_3, are considered as bases. The Brønsted and Lewis concepts of a base are practically equivalent, the concepts of acidity are different, however.

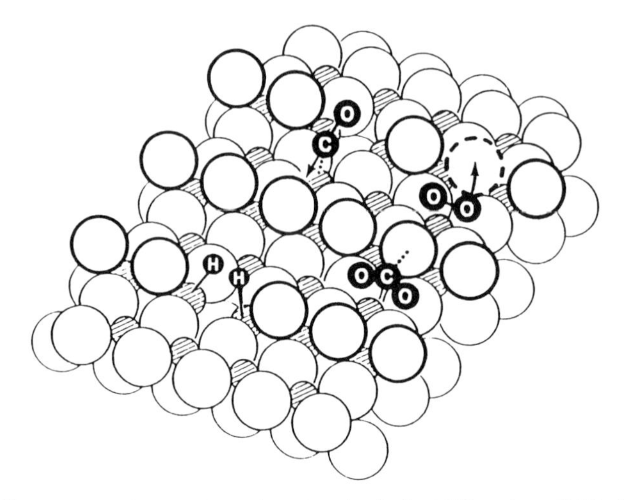

Figure 3.7. Oxygen vacancies constitute adsorption sites in the (110) surface of TiO_2 (from Göpel *et al.*, 1983).

The reactivity of an oxide surface depends on the extent to which the coordination of the surface ions is unsaturated. The densely packed and stable surfaces possess the lowest degree of coordinative unsaturation and are, in general, unreactive. Defects, in particular those associated with oxygen vacancies, expose cations whose coordination is highly unsaturated. Such sites may give rise to strong adsorption.

Figure 3.7 schematically illustrates the adsorption of several molecules on a defect surface of TiO_2. Oxygen vacancies serve as the adsorption sites. Dissociation of O_2 on a defect produces oxygen atoms which locally restore the surface structure by filling vacancies. Molecules such as CO and H_2 do not adsorb on the perfect TiO_2 (110) surface, but when defects are present, H_2 dissociates at temperatures as low as 80 K.

Characteristic of transition metal oxides is that the cation may have more than one oxidation state. Iron, for example, has three stable oxides, FeO, Fe_3O_4, and Fe_2O_3, and several oxyhydroxides, such as FeOOH in a number of different structures, as well. Surfaces of oxides with the cations in lower oxidation state are generally more reactive than those with the cation in its highest oxidation state. In addition, these ions can participate in reactions that involve changes in valence state.

Zeolites form a special class of oxides, which are catalytically active owing to the presence of acid sites. Zeolites are crystalline alumina-silicates with micropores defined by the crystallographic structure of the solid (Figure 3.8). Micropore dimensions may vary and can be of the order of the size of an organic molecule.

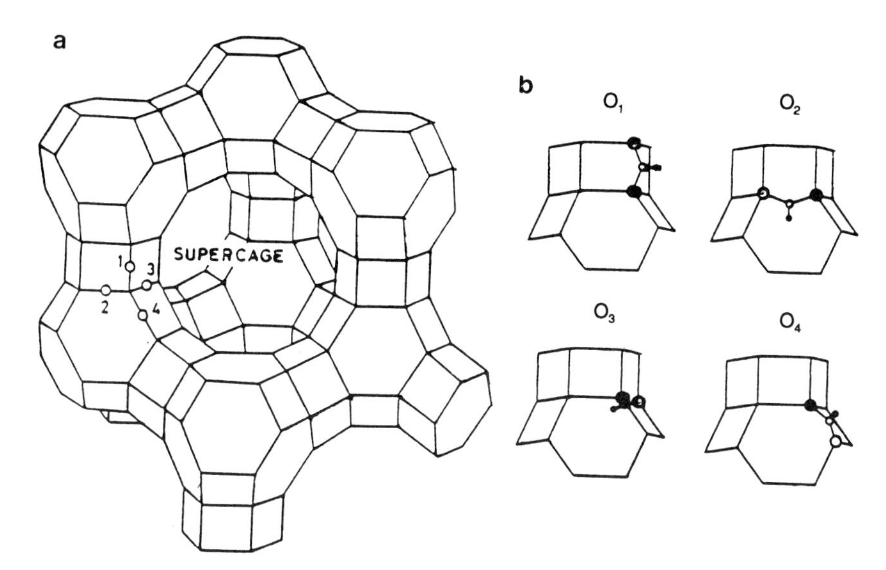

Figure 3.8. (a) Structure of faujasite, with four different O atoms in tetrahedral coordination to Si and Al; (b) Brønsted acid OH groups, bridging between Si and Al sites.

Protons attached to the oxygen atoms of the zeolite lattice are strong Brønsted acids and constitute the sites where hydrocarbons or basic molecules adsorb (Figure 3.8).

The following will describe the adsorption of a number of common molecules in catalytic reactions on oxide surfaces.

Water possesses a large dipole moment, as well as a lone pair of electrons on the oxygen. It is a molecule with good donor properties which adsorbs on the Lewis-acidic M^{n+}-cations of the surface. The dissociation into H^+ and OH^- occurs readily on almost all oxide surfaces, particularly when defects are present

$$\begin{array}{c} H_2O \\ -M^{\delta+}-O^{\delta-}-M^{\delta+}- \end{array} \longrightarrow \begin{array}{c} OH^{\delta-}\;\; H^{\delta+} \\ |\qquad | \\ -M^{\delta+}-O^{\delta-}-M^{\delta+}- \end{array} \tag{3.22}$$

(note that δ has only symbolic significance; it is not implied that all centers possess the same amount of charge). Two types of hydroxyl groups are present, the terminal OH on the cation has basic character (in aqueous solutions of low pH it will take up a proton), while the bridging OH between the two cations is a Brønsted acid because it can release a proton. In general, it is safe to assume that the surfaces of oxide catalysts, which have been exposed to air, will be saturated with hydroxyl groups and physisorbed water.

Hydrogen, being a non-polar molecule with weak donor-acceptor properties, interacts only weakly with oxide surfaces. The interaction depends strongly on the type of sites the surface exposes. As the H–H bond is strong, dissociation occurs only on the most reactive surfaces. Splitting is thought to occur heterolytically

$$\underset{\text{—}M^{\delta+}\text{—}O^{\delta-}\text{—}}{H_2} \longrightarrow \text{—}\overset{H^-}{\underset{|}{M^{\delta+}}}\text{—}\overset{H^+}{\underset{|}{O^{\delta-}}}\text{—} \qquad (3.23)$$

Oxygen adsorbs weakly on stoichiometric surfaces, but more strongly on defects and on oxides that have been partially reduced. Several adsorbed oxygen species exist: molecular oxygen, in the form of O_2, O_2^- or, less common, O_2^{2-}, and atomic oxygen in the form of O, O^- or O^{2-}. In addition, O^{2-} ions of the lattice need to be taken into account in catalytic reactions. The dissociation of molecular oxygen species requires two adjacent vacancies, the diffusion of oxygen species or of lattice vacancies make it an activated process. Nitrous oxide (N_2O) dissociates at low temperatures and represents a convenient source of atomic oxygen in surface studies.

Carbon monoxide has weak donor properties and adsorbs relatively weakly on coordinatively unsaturated cations (Figure 3.7). If the surface cations are in a lower oxidation state, CO adsorption may be stronger depending on the extent that back donation of electrons from the cation to the antibonding orbitals of the CO takes place. Dissociation of CO is highly unlikely. On the other hand, adsorbed CO may disrupt M–O bonds in the surface to form an M–CO_2 complex which, when the temperature is raised, desorbs as CO_2, leaving an oxygen vacancy behind. Other adsorption states exist as well. On fully oxidized surfaces, carbonates may form, while reaction with a surface hydroxyl group may yield a formate species. Carbon dioxide adsorbs on perfect surfaces in the form of a carbonate.

Nitrogen oxide (NO) interacts strongly with the coordinatively unsaturated cations of oxide surfaces than CO does. Characteristic of NO is the formation of dinitrosyl species, i.e., two NO molecules attached to the same metal ion, on many oxide surfaces, such as those of the catalytically relevant oxides of molybdenum, tungsten, iron, cobalt, and chromium.

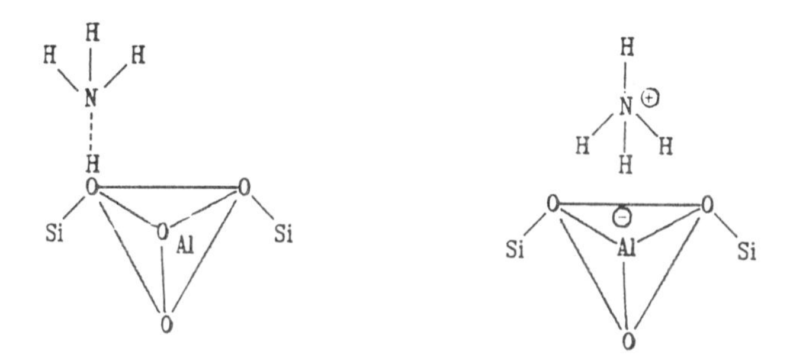

Figure 3.9. Adsorption of NH_3 followed by protonation on the acid site of a zeolite.

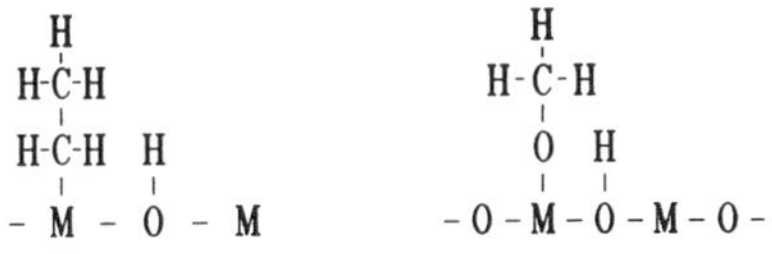

```
  H                      H
H-C-H                  H-C-H
  |                      |
H-C-H  H                 O  H
  |    |                 |  |
- M -  O - M       - O - M - O - M - O -
```

Figure 3.10. Adsorption modes of hydrocarbons and alcohols on oxide surfaces.

Ammonia (NH_3) possesses a lone pair of electrons on the nitrogen with which it adsorbs to the Lewis acidic, coordinatively unsaturated cations of the oxide. If hydroxyls possessing Brønsted acidity are present, as in zeolites, ammonia may become adsorbed as NH_4^+ (Figure 3.9). Hydrogen bonding with oxide anions of the surface is a third mode of bonding. However, this type of bonding is weak. Nitrogen (N_2) shows hardly any interaction with oxide surfaces beyond physisorption at very low temperatures.

Hydrocarbons and alcohols A general feature of the interaction between oxide surfaces and hydrocarbons is the facile breaking of C–H or O–H bonds while C–C bonds stay intact. For example, methane dissociates on several oxides to form hydroxyl and a methyl or methoxy groups (depending on whether the CH groups bond to the metal or the oxygen ion), while methanol readily decomposes to methoxy and hydroxyl groups (Figure 3.10). Other modes of bonding become available if the hydrocarbon has unsaturated bonds. Alkenes and alkynes are adsorbed molecularly in the form of a π-complex, which gives a relatively weak interaction or dissociatively. In the latter case, propylene forms the π-allyl anion

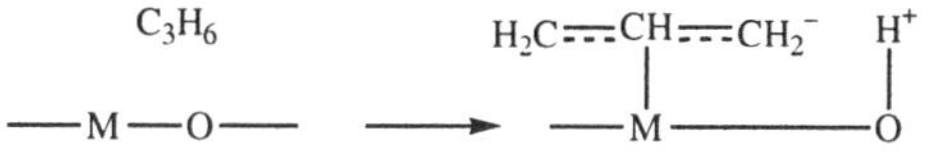

$$\underset{\text{—M—O—}}{C_3H_6} \longrightarrow \underset{\text{—M————O}}{H_2C\!\cdots\!CH\!\cdots\!CH_2^- \quad H^+} \tag{3.24}$$

while adsorption of propyne may produce either methyl acetylide ($CH_3–C\equiv C^-$) or the propagyl ion ($CH_2=C=CH^-$).

3.2.2. Catalytic Reactions on Oxides

Catalytic reaction mechanisms on oxide catalysts often involve oxygen atoms from the lattice ending up in product molecules. Multicomponent oxides are used as selective oxidation catalysts.

A consequence of the coordinative unsaturation at oxide surfaces is that the oxygen anions in the surface have weaker bonds with the underlying lattice than do the fully coordinated oxygen ions of the interior. As a result, lattice oxygens at the surface may participate in oxidation reactions, where they become incorporated in product molecules. This type of mechanism, named after Mars and Van Krevelen,

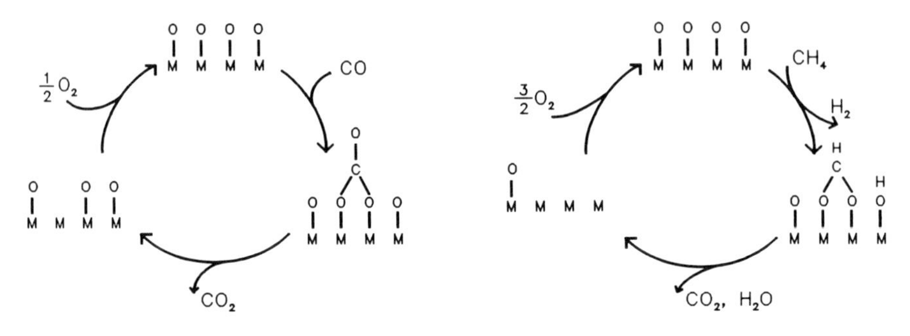

Figure 3.11. The catalytic reaction cycles for the total oxidation of (a) CO and (b) CH_4 on oxide catalysts.

is illustrated with the total oxidation of CO and hydrocarbons, which are relevant in environmental pollution control.

Figure 3.11a schematically shows the reactions involved in the oxidation of CO to CO_2 on a metal oxide surface, such as that of ZnO. The first step is the adsorption of CO in the form of a surface carbonate, as observable in infrared spectroscopy. Next CO_2 desorbs from the surface, leaving two oxygen vacancies behind. The cycle is completed by dissociative adsorption of oxygen, which produces oxygen atoms to fill the vacancies. According to Boreskov, such a stepwise mechanism prevails at higher temperatures (i.e., above 525 K on CuO), where the rates of reduction and reoxidation of the oxide surface are comparable to the overall rate of the CO oxidation. At lower temperatures, however, reduction and reoxidation slow down, whereas the rate of reaction decreases considerably. This is why desorption of CO_2 and reoxidation of the surface are believed to occur simultaneously in a concerted reaction step.

The total oxidation of hydrocarbons is believed to proceed through carboxylate intermediates. Figure 3.11b schematically illustrates this for the oxidation of methane: Adsorption of methane produces formate and hydroxyl groups on the surface, which decompose to CO_2 and H_2O, while dissociative adsorption of O_2 reoxidizes the surface. Also, in this case a stepwise mechanism is thought to operate at high temperatures, whereas desorption of CO_2 and H_2O and restoration of the oxidic surface occur in a concerted mechanism at lower temperatures. Interestingly, experiments with isotopically labeled $^{18}O_2$ have indicated that when a surface containing formate and hydroxyls is exposed to $^{18}O_2$, the latter does not appear in the products, but is used exclusively for reoxidation of the surface.

Where total oxidation is relevant for cleaning industrial exhaust gases, partial or selective oxidation is of great importance for the production of chemicals. The oxides used as catalysts are often multicomponent systems. The $(Bi_2O_3)_x(MoO_3)_y$ system is of practical use in the oxidation of propylene to acroleine

$$CH_2 = CH - CH_3 + O_2 \rightarrow CH_2 = CH - CHO + H_2O \qquad (3.25)$$

and related systems based on iron and antimony oxide are used in the catalytic ammoxidation to produce acrylonitrile

$$CH_2 = CH - CH_3 + \frac{3}{2} O_2 + NH_3 \rightarrow CH_2 = CH - C \equiv N + 3\, H_2O \qquad (3.26)$$

Acrolein and acrylonitrile are important intermediates for polymer production.

Selective and non-selective oxidation reactions nearly always require generation of atomic oxygen on the catalyst surface. The chemical properties and reactivity of these oxygen atoms differ widely, depending on the composition of the catalyst. The strength of the bond between oxygen and the catalyst is of great importance. Different reactions require oxygen atoms of different reactivity. According to Gates *et al.* (1979), the oxidation of an aromatic molecule needs a relatively weakly bound oxygen, while the oxidation of methanol needs oxygen that is more strongly held in order to prevent total oxidation. Olefins present an intermediate case. This implies that a catalyst for selective oxidation of aromatics would lead to total combustion if applied in oxidation reactions of olefins.

It appears that in the $(Bi_2O_3)_x(MoO_3)_y$ system the site for oxygen dissociation is different from that for olefin activation. The oxygen atoms generated upon dissociation have a high mobility and travel through the oxide to the olefin oxidation site. The reactive oxygen atoms are nucleophilic. They react with an allylic intermediate of propylene coordinated to a Mo ion, while hydrogen abstraction is catalyzed by Bi. Reaction networks of selective catalytic oxidations are highly complex and not well understood for most of the reactions. Kinetic measurements indicate that the rate of reaction for the $(Bi_2O_3)_x(MoO_3)_y$ catalyzed oxidation of propylene to acroleine (3.25) has a first order in propylene, zero order in oxygen, and a negative order in acroleine, indicating that acroleine adsorbs rather strongly on the catalyst. Refer to Gates *et al.* (1979) for more information on selective oxidations.

Multicomponent $(PO_4)_x(V_2O_5)_y$ catalysts are applied in processes to produce maleic anhydride and phthalic anhydride intermediates for polymers. In the classical industrial process, maleic anhydride is the product from benzene and phthalic anhydride the product from naphthalene. Vanadium oxide-based catalysts have a stronger interaction with their substrates than molybdenum oxide-based catalysts. Whereas on MoO_3 catalysts the hydrocarbon skeleton remains intact, oxidation on vanadium oxides proceeds with rupture of carbon–carbon bonds and the total number of carbon atoms in the molecule is not maintained in the reaction products. A new development is the use of butane as a reactant in these processes.

Unlike ethylene epoxide, which is produced from ethylene and oxygen on a silver catalyst, the epoxide of propylene cannot be obtained by direct catalytic oxidation. The non-catalytic chlorohydrin route is still in use, although it is gradually being replaced by indirect oxidation routes using organic hydroperoxides. In the oxirane process, ethylbenzene or isobutane are first oxidized to the corresponding hydroperoxide in the gas phase. In a subsequent liquid-phase oxidation, the peroxide oxygen is transferred selectively to the double bond of the propylene to form the epoxide

$$\Phi-\underset{\underset{H}{|}}{\overset{\overset{CH_3}{|}}{C}}-O-O-H + CH_3-CH=CH_2 \longrightarrow CH_3-\underbrace{CH-CH_2}_{O} + \Phi-\underset{\underset{OH}{|}}{C}-CH_3 \quad (3.27)$$

The epoxide is used as a monomer for polymer production. The byproduct ethylbenzene alcohol can be dehydrated to styrene, also a monomer for the production of polymers. If isobutane is used, iso-butylhydroperoxide replaces ethylbenzene-hydroperoxide as the oxidant. The byproduct tert-butanol can be converted with methanol to an ether that is an important additive in new environmental friendly gasolines. Complexes of Mo, V, or Ti are used in homogeneous epoxidation catalysis, while heterogeneous TiO_2/SiO_2 catalysts can be used also. The active sites consist of a titanium ion with a fourfold coordination of oxygen in a tetrahedral geometry. Titanium acts essentially as a Lewis acid to activate the O–O bond in the hydroperoxide.

Tetrahedrally-coordinated cations are also encountered in the lattice of zeolites (Figure 3.8). The well-defined nature of the lattice, the variety of microchannel dimensions, and the many possibilities for isomorphous substitution make them ideal catalysts for the production of fine chemicals. A recent application is the oxidation of phenol to catechol by hydrogen peroxide using a titanium–silicalite catalyst with Ti ions in tetrahedral sites.

Ziegler–Natta type polymerization catalysts for the production of polypropylene and polyethylene operate at low temperatures. Modern processes use $TiCl_3$ dispersed on $MgCl_2$ particles. The reaction proceeds by formation of an organometallic complex of the growing polymer chain bonded to one vacancy around Ti and the alkene coordinated to another ligand position of Ti, analogous to the binding in coordination complexes. This is, however, one of the few examples of the use of a heterogeneous polymerization catalyst. The fine-tuning possibilities by ligand variation of organometallic complexes and the mild conditions to be used makes exploitation of homogeneous catalysts very useful.

An analogous situation exists in disproportionation catalysis. Reactions of the type $R_1–C=C–R_1 + R_2–C=C–R_2 \rightarrow 2\,R_1–C=C–R_2$ are performed at a low temperature over heterogeneous ReO_3 or MoO_3 catalysts dispersed on Al_2O_3. Again catalysis proceeds by formation of one-center coordination complexes that transform as:

$$
\begin{array}{c} R_1\text{—}C\text{=}C\text{—}R_1 \\ | \\ M\text{=}CHR_2 \end{array}
\longrightarrow
\begin{array}{c} R_1\text{—}C\text{—}H \\ \| \\ M \end{array}
\cdots\cdots
\begin{array}{c} H\text{—}C\text{—}R_1 \\ \| \\ H\text{—}C\text{—}R_2 \end{array}
\qquad (3.28)
$$

Alkenes coadsorbed on the surface of a transition metal can oligomerize, provided that the temperature is not too high. The reaction proceeds with low selectivity, however, as it competes with hydrogenolysis. Mono-center homogeneous catalysts have been developed that dimerize alkenes with high selectivity. Also the hydroformylation reaction, in which a CO molecule inserts into an alkene to produce an aldehyde, is catalyzed by transition metals with low selectivity. Very selective and active homogeneous catalysts based on a single metal center operate in this reaction, which is important for the production of detergents. Currently an increasing number of new chemical processes is designed to operate with homogeneous catalysts under mild conditions.

Important new catalysts have been developed for environmental applications such as the reduction of NO by NH_3 in industrial stackgases, in the presence of oxygen. This reaction is catalyzed using TiO_2/V_2O_5. Another relevant example of catalytic clean-up technology is the removal of SO_2 from stackgases by a reaction with H_2S to produce sulfur and H_2O. In modern refinery processes this reaction is catalyzed by Fe/SiO_2 or Al_2O_3 catalysts. The mechanism of this process is still not quite understood.

Oxides can be used as catalysts for hydrogenation and dehydrogenation reactions. An important dehydrogenation process is the production of styrene from ethylbenzene. Styrene is the monomer unit of the polystyrene polymer. The dehydrogenation reaction proceeds typically around 500°C and is catalyzed by promoted Fe_2O_3. Steam is added to suppress deactivation of the catalyst by gasification of unreactive carbonaceous deposits.

3.2.3. Solid Acid Catalysis

Zeolites are widely applied in the processing of oil. Acid-catalyzed reactions of hydrocarbons proceed through carbenium ion intermediates. Bifunctional catalysis combines the catalytic properties of metal particles and of zeolites.

Transforming crude oil to gasoline is the most important process catalyzed by solid acids, such as chlorine-treated clays and zeolites. The reaction is performed at high temperatures (500°C) but is unfortunately accompanied by significant coke formation, causing a short lifetime for the catalyst. In a riser–downer reactor the catalyst is exposed for a short time to the feed, and in a second step, carbon is removed by oxidative combustion.

Application of zeolites has significantly improved gasoline yield. The protons form carbenium or carbonium ions with reactant hydrocarbon molecules that convert to smaller molecules by subsequent carbon–carbon bond scission reactions. Also oligomerization and aromatization reactions of olefins are catalyzed. Zeolites suppress coke formation that deactivates the catalyst. The finite size of the zeolite micropores prevents formation of large condensed aromatic molecules resulting from oligomerization reactions. Such large molecules do not fit in the zeolite pore system. This shape-selective effect results in a significantly improved carbon efficiency in the product and a longer lifetime for the catalyst. The economic success of zeolite applications has generated a large research effort to synthesize and discover new, improved zeolites. Currently, nearly 80 different structures are known with a widely varying composition. Whereas the wide-pore zeolite Y is applied in catalytic cracking, medium-pore zeolites are used as hydroisomerization catalyst, dewaxing, or alkylation catalysts.

In catalytic dewaxing, linear alkanes are separated from branched hydrocarbons by cracking the molecules over zeolites with micropores that access linear alkanes only, but not branched molecules. Branched alkanes are desired as high octane gasoline components. A low reaction temperature (200° C) is preferred for the isomerization reaction because isomerization is an exothermic reaction. In this reaction linear paraffins are isomerized and more branched molecules are produced.

Conventional cracking catalysts operate at a temperature of ~ 500–600°C. The difficult step is the generating carbenium ions from alkanes. This occurs through a reaction sequence that initially proceeds with formation of carbonium ions:

$$H^+ + \text{alkane} \rightarrow [C_nH_{2n+3}]^+ \tag{3.29}$$

The unstable carbonium ion decomposes rapidly to a carbenium ion:

$$[C_nH_{2n+3}]^+ \begin{cases} \nearrow [C_nH_{2n+1}]^+ + H_2 \\ \searrow [C_{n-x}H_{2n-2x+1}]^+ + C_xH_{2x} \end{cases} \tag{3.30}$$

Once the reaction has been initiated, the reaction is propagated by creating new carbenium ions by hydride transfer to product ions:

$$C_nH_{2n+2} + [C_mH_{2m+1}]^+_{ads} \rightarrow C_mH_{2m+2} + [C_nH_{2n+1}]^+_{ads} \tag{3.31}$$

This reaction requires a considerably lower activation energy than carbonium ion formation does. The carbenium ions undergo cracking or isomerization in subsequent reactions.

Solid acids are highly active alkylation catalysts. Alkylation, the introduction of an alkyl group into a molecule, is used to produce highly branched alkanes which are added to gasoline in order to improve the octane rating. Around half of the worldwide production of isobutene and n-butenes is used for this purpose. Methyl *tert*-butyl ether is a gasoline additive of rapidly growing importance. It is manufactured in the liquid phase from iso-butene and methanol using a Brønsted-acid catalyst, according to the reaction

$$(H_3C)_2C{=}CH_2 + CH_3OH \overset{H^+}{\rightleftharpoons} (H_3C)_2C(CH_3)(OCH_3) \qquad (3.32)$$

Brønsted acidic zeolites represent active alkylation catalysts in heterogeneous processes. An example is the alkylation of benzene with ethylene to ethylbenzene, which is the precursor for styrene. Conventionally this reaction has been catalyzed by Friedel–Crafts catalysts such as $AlCl_3$ in the liquid phase, using HCl as a cocatalyst. The process, however, consumes relatively large amounts of $AlCl_3$, requires costly separations and a reactor that is sufficiently resistant against corrosion. The gas-phase ethylation over solid acids such as SiO_2-Al_2O_3 or modified ZSM-5 zeolites is much more environmentally friendly.

Another important alkylation process is that of iso-butane and propylene. The resulting alkylate is used as kerosene for airplanes. Currently HF is used in a low temperature process. Obviously, a large incentive also exists to replace this corrosive, environmentally hostile catalyst by a zeolite.

The application of the zeolite, ZSM-5, to convert methanol to synthetic fuels represents a subject of recent interest (Figure 3.12a). Since methanol is produced from synthesis gas, the process can be used to convert natural gas to gasoline. Hydrocarbons beyond C_{10} are hardly formed, as these do not fit in the ZSM-5 pore system. Methanol first reacts to dimethyl ether which subsequently dehydrates to form ethylene. Oligomerization reactions lead to higher olefins, aromatics and hydrocarbons in the gasoline range. The medium-pore ZSM-5 zeolite appears to be an excellent and stable catalyst for this reaction; application of zeolites with smaller pores gives mainly small olefins. Figure 3.12b shows the product distribution as a function of contact time between catalyst and feed, illustrating that the reaction can be tuned toward production of light olefins or to gasoline.

The combination of metal and zeolite in one catalyst system offers attractive possibilities for isomerization reactions, for example in the reforming of naphta. The advantage is caused by the ease at which metals catalyze (de)hydrogenation reactions. Whereas carbenium ions on zeolites form readily by protonation of alkenes, the formation of carbonium ions from alkanes requires a high activation energy (see also Section 6.4.3) Because transition metals possess excellent activity for the dehydrogenation of alkanes (see also Section 3.1.2.3), the isomerization of

a

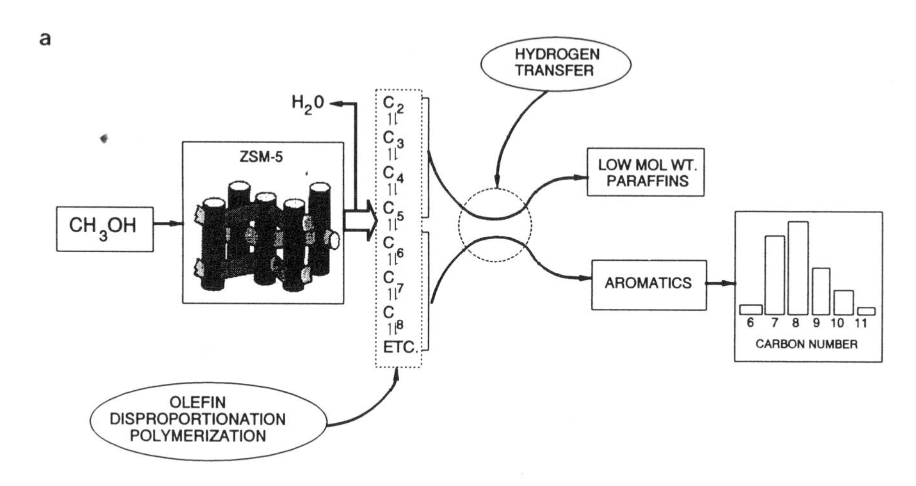

b

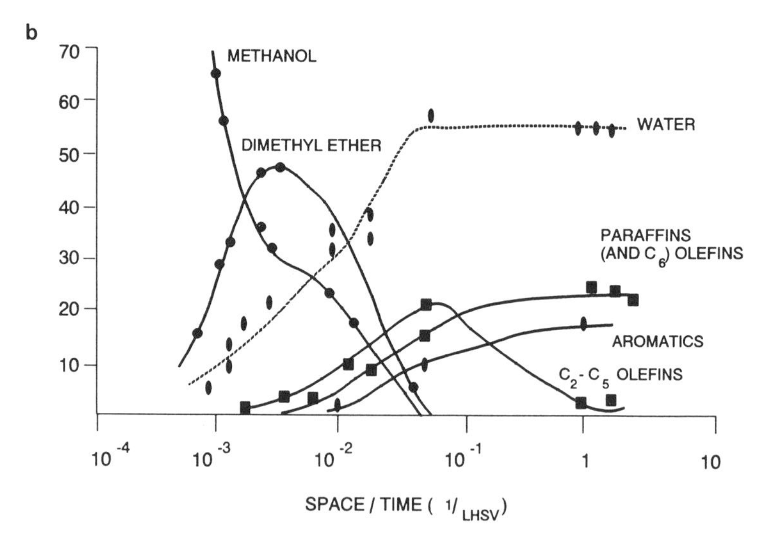

Figure 3.12. (a) The zeolite ZSM-5 catalyzes the reaction of methanol to hydrocarbons in the gasoline range; (b) Product distribution as a function of contact time between catalyst and feed (after Chang, 1983).

alkanes is carried out with bifunctional catalysts, consisting of small metal particles supported on the zeolite.

The following example illustrates how the metal and the acidic function cooperate in the isomerization of n-pentane to iso-pentane. The first step is the dehydrogenation of the alkane on the metal

$$n\text{-}C_5H_{12} \underset{Pt}{\overset{K_1}{\leftrightarrows}} n\text{-}C_5H_{10} + H_2 \tag{3.33}$$

The thus formed n-pentene adsorbs on an acid site of the support, where it reacts to the isomer iso-pentene

$$n\text{-}C_5H_{10} + H^+_{acid} \overset{K_2}{\leftrightarrows} [n\text{-}C_5H_{11}]^+_{ads} \tag{3.34}$$

$$[n\text{-}C_5H_{11}]^+_{ads} \overset{k_3}{\leftrightarrows} [i\text{-}C_5H_{11}]^+_{ads} \tag{3.35}$$

$$[i\text{-}C_5H_{11}]^+_{ads} \overset{K_4}{\leftrightarrows} [i\text{-}C_5H_{10}] + H^+_{acid} \tag{3.36}$$

Finally, the iso-pentene is hydrogenated to iso-pentane on the metal

$$i\text{-}C_5H_{10} + H_2 \underset{Pt}{\overset{K_5}{\leftrightarrows}} i\text{-}C_5H_{12} \tag{3.37}$$

If the catalyst contains sufficient platinum to allow the hydrogenation/dehydrogenation steps to be in equilibrium, using the actual isomerization as the rate-limiting step, the following expression can be obtained for the rate if the backward reaction is ignored

$$r = \frac{N_{acid}k_3K_1K_2[n\text{-}C_5H_{12}]/[H_2]}{1 + K_1K_2[n\text{-}C_2H_{12}]/[H_2]} \tag{3.38}$$

which depends only on the ratio of the partial pressures of hydrogen and n-pentane. Support for the mechanism is provided by the fact that the rate of n-pentene isomerization on a platinum-free catalyst is very similar to that of the above reaction.

Bifunctional processes are used commercially to convert C_5 and C_6 alkanes to isomerized products. An additional benefit of the bifunctional catalyst is that the concentration of alkenes in the gas phase remains low, which suppresses consecutive cracking and oligomerization reactions. The latter leads to coke formation and eventually deactivates the catalyst.

This book can present only a few illustrative highlights of zeolite catalysis. Please refer to the review edited by van Bekkum *et al.* (1991).

3.3. CATALYSIS BY SULFIDES

Catalysts based on MoS_2 are used to remove sulfur and nitrogen from organic molecules in hydrodesulfurization and hydrodenitrogenation processes.

Sulfides of molybdenum and tungsten, often promoted with cobalt or nickel and supported on alumina, are used for the hydrotreating of heavy oil fractions. Although sulfidic catalysts have been used for desulfurization since this century's beginning, much less is known of adsorption of gases on sulfides than on metals or even oxides. This discussion is limited to the sulfide MoS_2.

Molybdenum disulfide has a layered structure consisting of Mo with a formal oxidation state of 4+, trigonally coordinated by sulfur anions. The basal plane of this structure is virtually unreactive, all activity is associated with the edges of the slabs (Figure 3.13). These are the sites where adsorption takes place. O_2, CO and NO are among the gases which have been shown to adsorb at these sites. Hydrogen is easily activated by sulfides, and dissociates readily under reactions conditions.

Catalytic desulfurization and denitrogenation have been developed in the beginning of this century to clean the sulfur- and nitrogen-rich oils provided by coal

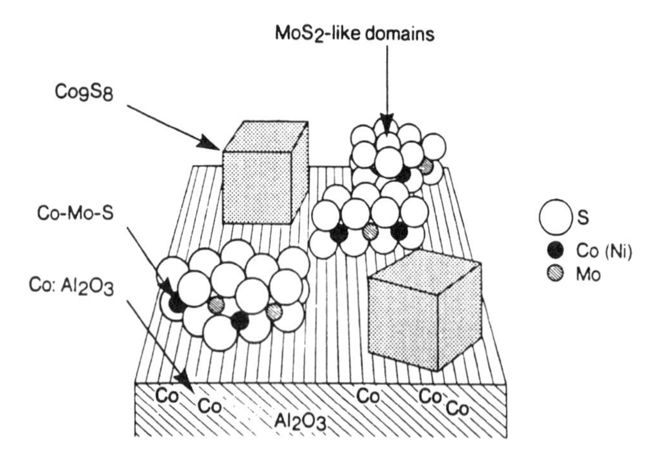

Figure 3.13. Schematic representation of the sulfided Co-Mo/Al_2O_3 catalysts, showing MoS_2 slabs decorated with cobalt on the edges (from Topsøe and Clausen, 1984).

liquefaction. Currently these processes are widely applied in the oil industry to treat heavy feedstocks. Hydrogen disulfide and ammonia are formed by hydrogenolysis of C–S and C–N bonds in large molecules. The most commonly applied catalysts are based on molybdenum sulfide promoted with nickel or cobalt, which are also present in the sulfided state.

Fundamental studies of hydrodesulfurization kinetics often use thiophene (C_4H_4S) to represent the type of sulfur bonding present in larger molecules. A simplified scheme for the reaction is illustrated in Figure 3.14 and corresponds to

$$\begin{aligned} T + \square &\leftrightarrows T_{ads} \\ H_2 + 2* &\leftrightarrows 2H_{ads} \\ T_{ads} + 2H_{ads} &\leftrightarrows B + S_{ads} + 2* \\ S_{ads} + 2H_{ads} &\leftrightarrows H_2S + 2* + \square \end{aligned} \tag{3.39}$$

where T stands for thiophene and B for butadiene, believed to be the primary product in the absence of secondary hydrogenations. Note that in (3.39) hydrogen adsorbs on different sites than thiophene, which has been assumed to coordinate to sulfur vacancies at the edges of a MoS_2 particle. Although the actual mechanism of the hydrodesulfurization reaction is still under debate, we have deliberately chosen the mechanism of (3.39) because it illustrates the kinetics of processes involving two kinds of sites. As a consequence, two site balances are obtained

$$\theta_\square + \theta_T + \theta_S = 1; \quad \theta_H + \theta_* = 1 \tag{3.40}$$

Assuming the breaking of the S–C bonds to be rate-limiting the rate obtained

$$\begin{aligned} r = -V\frac{d[T]}{dt} &= N_\square k_3 \theta_T \theta_H^2 \\ &= \frac{N_\square k_3 K_1 K_2\,[T][H_2]}{(1 + K_1[T] + K_4^{-1}[H_2S])(1 + K_2^{1/2}[H_2]^{1/2})^2} \end{aligned} \tag{3.41}$$

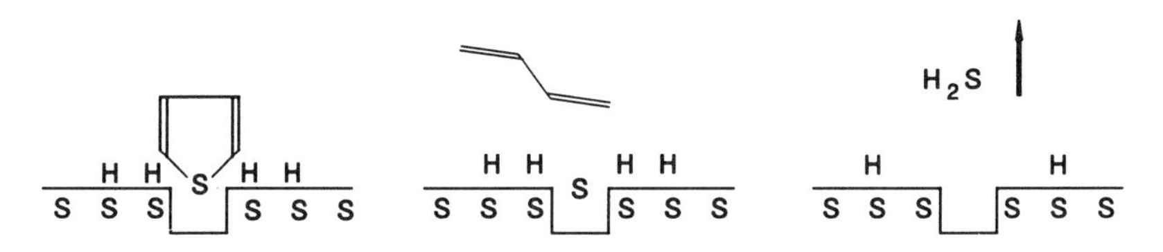

Figure 3.14. Simple reaction scheme for the desulfurization of thiophene on a sulfidic catalyst.

As the adsorption of hydrogen is rather weak under reaction conditions, the term $(1+(K_2H_2)^{1/2})^2$ in the denominator of (3.41) may be omitted. The rate expression shows that the reaction is suppressed by H_2S. Catalysts that are less sensitive to H_2S poisoning are usually the most active. Although the active Co-Mo/Al_2O_3 catalyst has extensively been studied and detailed models of its structure have been proposed (Figure 3.13), the molecular mechanism by which the catalyst operates is still not completely understood.

COLLISION AND REACTION-RATE THEORY

This chapter will show how equilibrium concentrations and reaction rate constants depend on microscopic properties of the reacting molecules, such as bond energies, vibrational frequencies, and rotational moments, which can be studied spectroscopically. A short introduction to the necessary statistical-thermodynamical background is given. Finally the theoretical rate expressions are used to discuss the rate constants of some elementary surface reactions.

4.1. MICROSCOPIC THEORY AND THERMODYNAMICS

> *The energy of a system in equilibrium is exponentionally distributed over its degrees of freedom according to the Boltzmann distribution function. Thermodynamic quantities can be calculated from the partition functions of a system. We derive expressions for the chemical potential, energy and entropy. These expressions relate microscopic molecular properties to the thermodynamic properties of the macroscopic system.*

The second law of thermodynamics states that, *under equilibrium conditions, the entropy is at maximum*. This law can be used to determine equilibrium expressions for thermodynamic properties. However particular constraints also have to be satisfied. In a closed system of interacting molecules at equilibrium, entropy is maximized under the condition that the total number of particles as well as the total energy are constant. As will be shown this determines the energy distribution function of the particles in the system. The entropy is computed from W, the number

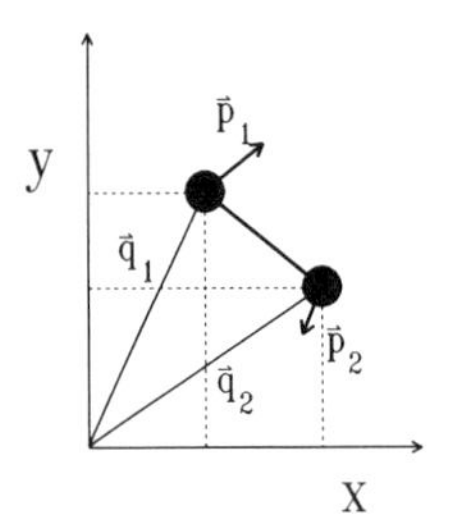

Figure 4.1. A diatomic molecule in phase space. The position and motion of the molecule are represented by the point with coordinates ($q_{1x}, q_{1y}, q_{1z}\ q_{2x}, q_{2y}, q_{2z}\ p_{1x}, p_{1y}, p_{1z}\ p_{2x}, p_{2y}, p_{2z}$) in a 12-dimensional phase space. For clarity, the z-components are not indicated.

of ways particles can be distributed over the different states of a system. A derivation of W will be started with to arrive at the relation between W and the entropy.†

Suppose there are N identical molecules, each consisting of f atoms, in a box of volume V. According to classical mechanics, the dynamics of each atom is defined by its 3 coordinates (e.g., x, y, z) and 3 momenta (e.g., p_x, p_y, p_z) (Figure 4.1). A molecule consisting of k atoms defines a $6k$ dimensional phase space. The $3k$ spatial coordinates will be labeled as $q_1, q_2, \ldots, q_{3k}$ and the $3k$ momenta as $p_1, p_2, \ldots, p_{3k}$. Three of the qs and ps can be chosen to correspond to the coordinates of the center of mass

$$\vec{R}_{cm} = \frac{\sum_i m_i \vec{q}_i}{\sum_i m_i} \tag{4.1}$$

$$\vec{P}_{cm} = \sum_i \vec{p}_i \tag{4.2}$$

Thus, in classical mechanics, a point in phase space corresponds uniquely with a well defined position and motion of a molecule. This is different in quantum mechanics, where position and momentum of a particle cannot be determined with arbitrary accuracy at the same time. The accuracy is restricted by the Heisenberg uncertainty principle

$$d\vec{p}\, d\vec{q} \geq h \tag{4.3}$$

in which h is Planck's constant. Therefore the molecule is confined to a region in phase space of the order of h^{3k}. It needs to be determined how many different ways

†Sections 4.1, 4.2.1, 4.2.2, and 4.4.1 closely follow Rice, 1967.

representative points can be distributed in the phase space, when the total number of particles as well as the total energy is fixed.

Each molecule has a particular energy-content, made up of translational, vibrational and rotational energy. The dimension of phase space is energy × time. So for each moment in time, phase space can be considered to consist of energy distributed over the 6k-1 remaining coordinates.

Divide the phase space into cells, each of which should contain a large number C of energy values, (e.g., translational energies). Suppose that the number of molecules of energy ε_1 is n_1, of ε_2 is n_2, etc. The total number of particles is constant

$$\sum_i n_i = N \tag{4.4}$$

The molecules are considered to be equivalent, so in one cell the number of ways in which the molecules can be distributed over the number of energy levels C is given by

$$\frac{C^{n_i}}{n_i!} \tag{4.5}$$

Each molecule has C possibilities to choose its energy. The factor $n_i!$ prevents double counting because of equivalence of the molecules. The total number of ways of distributing the molecules in their respective cells becomes

$$\frac{C^{n_1}}{n_1!} \cdot \frac{C^{n_2}}{n_2!} \cdots \frac{C^{n_i}}{n_i!} \cdots = \frac{C^N}{n_1!n_2! \ldots n_i! \ldots} = \frac{C^n}{\prod_i n_i!} \tag{4.6}$$

To find W, the number of ways particles can be distributed over all states of the system, sum over all different ways in which the molecules can be distributed over the cells

$$W = \sum_{(n_i)} \frac{C^N}{\prod_i n_i!} \tag{4.7}$$

The entropy can now be defined as a logarithmic function of the probability distribution

$$S = k \ln W \tag{4.8}$$

in which k is Boltzmann's constant. This definition represents a choice which is justified solely because it leads to the well known and highly useful thermodynamic relations, as will be shown. The maximum value of ln W for large values of N is of interest because this determines the state of maximum entropy in a macroscopic system.

Using Stirling's theorem

$$\lim_{N\to\infty} \ln N! = N \ln N - N \tag{4.9}$$

The entropy has an extremum when

$$0 = \delta \ln W = \delta \ln \frac{C^N}{n_1!n_2!\ldots.n_i!} \tag{4.10}$$

$$= -\sum_i \delta \ln n_i! + \delta N \ln C \tag{4.11}$$

Using Stirling's theorem this becomes

$$\delta n_1(\ln n_1 - \ln C) + \delta n_2(\ln n_2 - \ln C)\ldots \delta n_i(\ln n_i - \ln C) = 0 \tag{4.12}$$

Conservation of the total number of particles gives

$$0 = \delta N = \sum_i \delta n_i \tag{4.13}$$

and conservation of the total energy gives

$$0 = \delta E = \sum_i \varepsilon_i \delta n_i \tag{4.14}$$

The expressions (4.12)–(4.14) are satisfied simultaneously, but they are not independent. In order to avoid mathematical problems, constants α and β (called Lagrange multipliers) are introduced and combined with (4.12)–(4.14) into one condition

$$\delta n_1(\ln n_1 - \alpha - \beta\varepsilon_1) + \delta n_2(\ln n_2 - \alpha - \beta\varepsilon_2) + \ldots$$

$$\ldots \delta n_i(\ln n_i - \alpha - \beta\varepsilon_i) + \ln C \sum_i \delta n_i = 0 \tag{4.15}$$

α and β will now be determined. Relation (4.15) is satisfied when the coefficients of δn_i are zero. The solution of (4.15) thus becomes

$$\ln n_i - \alpha - \beta\varepsilon_i - \ln C = 0 \tag{4.16}$$

or

$$n_i = Ce^{\alpha}e^{\beta\varepsilon_i} \tag{4.17}$$

Since

$$N = \sum_i n_i \tag{4.18}$$

$$N = Ce^{\alpha} \sum_i e^{\beta\varepsilon_i} \quad (a)$$

$$\frac{N}{C} e^{-\alpha} = \sum_i e^{\beta\varepsilon_i} \quad (b) \tag{4.19}$$

If $x_i = n_i/N$ is defined as the fraction of particles with energy ε_i, the constants α and C drop out, which leaves

$$x_i = \frac{n_i}{N} = \frac{e^{\beta\varepsilon_i}}{\sum_i e^{\beta\varepsilon_i}} \tag{4.20}$$

Expression (4.20) contains the important result that the energy distribution is an exponential function of ε_i. The constant β can be related to the temperature by computing S and E and using the thermodynamic relation

$$\frac{\partial S}{\partial E} = \frac{1}{T} \tag{4.21}$$

The entropy S follows from (4.7) and (4.8)

$$S = k \ln \frac{C^N}{n_1!n_2! \ldots n_i! \ldots} \tag{4.22}$$

The distribution function n_i is chosen such that it yields the maximum value of S. Stirling's theorem gives

$$S = kN \ln C - k \sum_i n_i \ln n_i + kN \tag{4.23}$$

Using (4.17) this becomes

$$S = -kN\alpha - k\beta E + kN \tag{4.24}$$

with

$$E = \sum_i n_i \varepsilon_i \tag{4.25}$$

According to (4.24)

$$\frac{\partial S}{\partial E} = -k\beta \tag{4.26}$$

Comparison with (4.21) gives

$$\beta = -\frac{1}{kT} \tag{4.27}$$

The distribution function derived is the Boltzmann distribution

$$x_i = \frac{e^{-\varepsilon_i/KT}}{\sum_i e^{-\varepsilon_i/kT}} \tag{4.28}$$

Substitution of relation (4.19b) eliminates α from (4.24) to give

$$S = \frac{E}{T} - kN\ln\frac{N}{C} + kN\ln\left(\sum_i e^{-\varepsilon_i/kT}\right) + kN \tag{4.29}$$

The Helmholtz energy is defined as

$$A = E - TS \tag{4.30}$$

and the chemical potential

$$\mu = \frac{\partial A}{\partial N} \tag{4.31}$$

so this gives

$$\mu = -kT\ln C\left(\sum_i e^{-\varepsilon_i/kT}\right) + kT\ln N \tag{4.32}$$

By letting the label i not just sum the energies in one cell, but all the states of phase space, the constant C equals 1 and expression (4.32) reduces to

$$\mu = -kT\ln\left(\sum_i e^{-\varepsilon_i/kT}\right) + kT\ln N \quad (a)$$
$$= \mu_0 + kT\ln N \quad (b) \qquad (4.33)$$

The first term on the right of (4.33a) is equal to the chemical potential per molecule μ_0, whereas the second term is the concentration-dependent part.

The partition function plays a key role in the theory of reaction rates. It is equal to the denominator of (4.28) and defined as

$$p.f. = \sum_i e^{-\varepsilon_i/kT} \qquad (4.34)$$

The summation concerns all energy levels. The relation between chemical potential and partition function becomes

$$e^{-\mu_0/KT} = p.f. \quad \text{or} \quad \mu_0 = -kT\ln p.f. \qquad (4.35)$$

Expression (4.35) is a key relation for chemical kinetics. The average energy per molecule, ε, follows from (4.34) by using the definition

TABLE 4.1. Partition Functions[a]

definition:	$pf = \sum_i e^{-\varepsilon_i/kT}$	(4.34)
energy:	$\varepsilon = kT^2 \frac{\partial}{\partial T} \ln pf$	(4.36)
chemical potential:	$\mu_o = -kT\ln pf$	(4.35)
entropy:	$s = \frac{\partial}{\partial T} kT\ln pf$	(4.37)
translation:	$pf_{trans} = l\frac{(2\pi mkT)^{1/2}}{h}$	(4.40)
vibration:	$pf_{vib} = \frac{1}{1-\exp(-h\nu/kT)}$	(4.69)
rotation:	$pf_{rot} = \frac{8\pi^2 I kT}{h^2}$	(4.74)
equilibrium constant:	$K = \prod_i (pf)^{\nu_i}$	(4.81)
N equivalent particles:	$pf_N = \frac{1}{N!}(pf)^N$	(4.43)

[a]Note that the partition function for translation is given per degree of freedom, i.e. for one dimension. The partition function for vibration is given with respect to the lowest vibrational level and not the bottom of the vibrational potential.

$$\varepsilon = \frac{\sum_i \varepsilon_i e^{-\varepsilon_i/kT}}{\sum_i e^{-\varepsilon_i/kT}} = kT^2 \frac{\partial}{\partial T} \ln p.f. \tag{4.36}$$

Because $\mu_0 = \varepsilon - Ts \rightarrow s = \frac{\varepsilon - \mu_0}{T}$ it follows that

$$s = kT \frac{\partial}{\partial T} \ln p.f. + k \ln p.f.$$

$$= \frac{\partial}{\partial T} kT \ln p.f. \tag{4.37}$$

Once the partition function is known, the average energy and entropy per molecule, as well as the chemical potential can be directly calculated. These relations together with expressions for the partition function are summarized in Table 4.1.

4.2. PARTITION FUNCTIONS

The partition function of a system plays a central role in statistical thermodynamics. The concept was first introduced by Boltzmann, who gave it the German name *Zustandssumme*, i.e., a sum over states. The partition function is an important tool because it enables the calculation of the energy and entropy of a molecule, as well as its equilibrium. Rate constants of reactions in which the molecule is involved can even be predicted. The only input for calculating the partition function is the molecule's set of characteristic energies, ε_i, as determined by spectroscopic measurements or by a quantum mechanical calculation. In the next section the entropy and energy of an ideal monoatomic gas and a diatomic molecule is computed.

4.2.1. Partition Function of an Ideal Monoatomic Gas

The energy per molecule ε and the entropy per molecule s in an ideal gas are calculated from the partition function.

To compute the partition function of a freely moving atom, the summation over i in (4.34) has to be replaced by an integration over the variables defining the energy of a free atom. The translational energy of an atom (or center of mass motion of a molecule) is

$$E(p) = \frac{p_x^2 + p_y^2 + p_z^2}{2m} = \frac{1}{2m} p^2 \tag{4.38}$$

The movement of the atom is defined by its momenta and spatial coordinates. These define what is known as the phase space. The minimum volume per representative point in phase space equals h^3, as follows from Heisenberg's uncertainty principle: $dx\, dp_x \geq h$. The normalized space density element becomes

$$\frac{1}{h^3}\, dp_x\, dp_y\, dp_z\, dx\, dy\, dz \tag{4.39}$$

In the definition of the partition function (4.34), the index i sums all elements in space. In the present case it is replaced by an integral over momenta and coordinates of the atom

$$\begin{aligned} p.f. = \sum_i e^{-\varepsilon_i/kT} &\equiv \frac{1}{h^3} \int\!\!\int\!\!\int\!\!\int\!\!\int\!\!\int dp_x\, dp_y\, dp_z\, dx\, dy\, dz\, e^{-p^2/2mkT} \\ &= V\left(\frac{1}{h}\int_{-\infty}^{+\infty} dp_x\, e^{-p_x^2/2mkT}\right)^3 \\ &= V\left(\frac{2\pi mkT}{h^2}\right)^{3/2} \end{aligned} \tag{4.40}$$

This is the value of the translational partition function in three dimensions. V is the volume available per atom. Thus the translational partition function increases with increasing mass and temperature and decreasing pressure (because the volume per molecule enlarges). Once the partition function is known, the energy and entropy per atom of the ideal gas can be computed. Substitution of (4.40) into (4.36) gives

$$\begin{aligned} \varepsilon &= kT^2 \frac{\partial}{\partial T} \ln\left(V\left(\frac{2\pi mkT}{h^2}\right)^{3/2}\right) \\ &= \frac{3}{2} kT \end{aligned} \tag{4.41}$$

The atom has three degrees of freedom along the x, y, and z coordinate. Per atom, each degree of freedom contributes $\frac{1}{2}kT$ to the energy. If the energy is expressed per mole of gas, the Boltzmann constant k has to be replaced by the gas constant R. The translational entropy contribution per atom in a gas of N atoms follows from (4.29), (4.40) and (4.9), (remember that $C = 1$)

$$s = \frac{\varepsilon}{T} + k \ln\left(\sum_i e^{-\varepsilon_i/kT}\right) - \frac{k}{N} \ln N!$$

$$= \frac{5}{2} k + k \ln\left(\frac{V}{N}\left(\frac{2\pi mkT}{h^2}\right)^{3/2}\right) \tag{4.42}$$

The translational entropy per atom depends on the number of atoms, the volume per atom, its mass, and the temperature, whereas the translational energy per atom in an ideal gas depends on the temperature only.

An alternative way to derive the entropy per atom in a gas of N atoms is to start with the partition function of N equivalent particles $p.f.\ (N)$

$$p.f._N = \frac{1}{N!} p.f.^N \tag{4.43}$$

Using the definition of the partition function (4.34), expression (4.33) follows. For noninteracting particles or, more generally, non-interacting degrees of freedom, the total partition function is the product of the partition functions corresponding to each set of non-interacting degrees of freedom, multiplied by a degeneracy factor. The term $1/N!$ is the degeneracy factor giving the configurational entropy contribution in expression (4.33). Instead of (4.35), there is now a relation between μ_N and $p.f.(N)$

$$e^{-\mu/kT} = p.f.(N) \tag{4.44}$$

4.2.2. Classical Partition Function of a Diatomic Molecule

The energy and entropy are calculated from the partition function of a classical diatomic molecule. The internal entropy of a molecule depends logarithmically on its effective volume. The effective length of a molecule depends reciprocally on the force constant of the bond.

A diatomic molecule has two atoms and hence six degrees of freedom. It is convenient to separate three degrees of freedom corresponding to the center of mass. The other degrees of freedom are then relative coordinates. In the present case the number of internal molecular degrees of freedom is

$$3f - 3 = 3 \tag{4.45}$$

The same holds for the internal momenta. Earlier in (4.1) the center of mass coordinates and momenta was defined. The center of mass coordinates are those of the vector

$$\vec{R}_{cm} = \frac{\sum_i m_i \vec{q}_i}{\sum_i m_i} = \frac{\sum_i m_i \vec{q}_i}{M_{cm}} \tag{4.46}$$

and the relative coordinates are those of the vector

$$\vec{r} = \vec{q}_1 - \vec{q}_2 \tag{4.47}$$

The energy of the molecule equals

$$E = \sum_i \frac{1}{2m_i} p_i^2 + V(\vec{q}_1 - \vec{q}_2) \tag{4.48}$$

in which the potential energy only depends on the relative coordinates. The kinetic energy becomes

$$\sum_i \frac{1}{2m_i} p_i^2 = \frac{1}{2M_{cm}} P_{cm}^2 + \frac{1}{2\mu} p_r^2 \tag{4.49}$$

By substitution it can be found that reduced mass and momentum are given by

$$\mu = \frac{m_1 m_2}{m_1 + m_2} \qquad (a)$$

$$\vec{p}_r = \mu \vec{r} \qquad (b) \tag{4.50}$$

where P_{cm} is the momentum related to the center of mass and p_r is the momentum related to the reduced mass. Using the center of mass and relative coordinates, the partition function becomes

$$p.f. = \iint d\vec{R}_{cm}\, d\vec{r} \iint d\vec{P}_{cm}\, d\vec{P}_r\, e^{-((P_{cm}^2/2M_{cm}) + (p_r^2/2\mu) + V(\vec{r}))/kT}$$

$$\int d\vec{R} = \iiint dX\, dY\, dZ,\ etc \tag{4.51}$$

and $\qquad P^2 = P_x^2 + P_y^2 + P_z^2$

The degrees of freedom corresponding to center of mass motion and relative motion do not interact, thus *p.f.* can be written as

$$p.f. = p.f._{cm} \cdot p.f._{\text{rel}} \tag{4.52}$$

The translational partition function of the center of mass is the same as that for a monoatomic gas (4.40) where the mass m is replaced by the center of mass, M_{cm}. The partition function of relative motion becomes

$$p.f._{\text{rel}} = \left(\int_{-\infty}^{+\infty} dp_r \, e^{-p_r^2/2\mu kT} \right)^3 4\pi \int_0^\infty dr \, r^2 \, e^{-V(r)/kT} \tag{4.53}$$

The factor 4π stems from the replacement of cartesian by polar coordinates and performing the integration over φ and Θ

$$dx\, dy\, dz = d(cos\ \Theta) d\varphi\ r^2\, dr \qquad -1 < d \cos \Theta < 1 \tag{4.54}$$

$$0 < \varphi < 2\pi$$

$$0 < r < \infty$$

The integral (4.53) can of course only be calculated if an explicit expression for the potential $V(\vec{r})$ is chosen. Therefore (4.53) will be evaluated for the case that the relative motion is due to vibration and $V(\vec{r})$ will be taken for the potential of the harmonic oscillator

$$V(r) = U_0 + k_F(r - r_e)^2 \tag{4.55}$$

in which r_e is the equilibrium distance and k_F the force constant between the atoms; U_0 is the minimum value of the potential and gives a constant contribution to the partition function. Using (4.55), the partition function becomes

$$p.f. = V \left(\frac{2\pi M_{cm} kT}{h^2} \right)^{3/2} \left(\frac{2\pi \mu kT}{h^2} \right)^{3/2} 4\pi r_e^2 \left(\frac{\pi kT}{k_F} \right)^{1/2} e^{-U_0/kT} \tag{4.56}$$

In (4.56) the volume occupied by the molecule is recognized. The next two terms are the partition functions due to the kinetic energy of the center of mass and to relative kinetic energy. The term

$$4\pi r_e^2 \left(\frac{\pi kT}{k_F} \right)^{1/2} \tag{4.57}$$

has the dimension of a volume. $(\pi kT/k_F)^{1/2}$ is the effective length of the diatomic molecule.

The next section will show that the resulting classical expressions for energy and entropy are valid as long $kT >> h\nu$, ν being the frequency of the diatomic molecule. The expression for the frequency is given by

$$\nu = \frac{1}{2\pi}\left(\frac{2k_F}{\mu}\right)^{1/2} \tag{4.58}$$

When a molecule has high frequencies, the quantum mechanical expressions for the energy have to be used. This will be discussed in the next section.

The classical expression for the energy and entropy of a diatomic molecule will be derived from (4.56). According to (4.36) the energy becomes

$$\varepsilon_{cl} = kT^2 \frac{\partial}{\partial T} \ln p.f.$$

$$= \frac{7}{2} kT + U_0 \tag{4.59}$$

The center of mass motion of the molecule contributes $3kT/2$. The extra contribution comes from the relative motion of the atoms and is due to vibration and rotation.

Similarly, the classical expression for the entropy is given by

$$s_d = \frac{9}{2} k + k \ln \frac{V}{N}\left(\frac{2\pi M_{cm} kT}{h^2}\right)^{3/2} + k \ln\left(4\pi r_e^2 \left(\frac{\pi kT}{k_F}\right)^{1/2}\right) + k \ln\left(\frac{2\pi\mu kT}{h^2}\right)^{3/2} \tag{4.60}$$

The second term in the right half of (4.60) is the contribution due to the center of mass-motion, the last due to the relative vibrational motion. The effective volume of the molecule is a cylinder with radius r_e and length $(\pi kT/k_F)^{1/2}$. The entropy increases the smaller the force constant k_F or the weaker the bond in the molecule.

4.2.3. The Quantum-Mechanical Partition Function of the Diatomic Molecule

The general expressions for the energy and the entropy of a diatomic molecule are given. At high frequencies (these are often important) use the quantum mechanical expressions. The temperature dependent enthalpy and entropy depend on the frequencies of the system. The equilibrium constant for dissociation is related to the product and quotient of partition functions. Usually the partition function due to the rotational degrees of motion is much larger than that of the vibrational motion.

According to quantum mechanics, the vibrational and rotational motion of a molecule is quantized. The energy of a molecule is written as a sum of three terms:

$$E^{n,j}_{\text{diat}} = \frac{P^2_{cm}}{2M_{cm}} + \left(n + \frac{1}{2}\right)h\nu + \frac{j(j+1)h^2}{8\pi^2 I} + U_0 \tag{4.61}$$

in which ν is the frequency of vibrational motion of the molecule, given by (4.58), while I is the moment of inertia (4.55)

$$I = \mu r_e^2 \tag{4.62}$$

Relation (4.61) can be understood by rewriting the relative kinetic energy as

$$\frac{p^2_{\text{rel}}}{2\mu} = \frac{p^2_{rel,x} + p^2_{rel,y} + p^2_{rel,z}}{2\mu} \tag{4.63a}$$

$$= \frac{p_r^2}{2\mu} + \frac{p_a^2}{2\mu r^2} \tag{4.63b}$$

$$pr = \mu\, \dot{r}$$

$$p_a = mr\,\dot{Q}$$

The first term is the relative kinetic energy, the second the centrifugal energy. The angular momentum is given by p_a.

The contribution

$$\frac{j(j+1)h^2}{8\pi^2 \mu r^2} \tag{4.64}$$

follows from quantization of the angular momentum

$$p_a \equiv j(j+1)\hbar^2 \tag{4.65}$$

$$j = 0, 1, 2, \cdots$$

The vibrational energy follows from quantization of the internal energy contribution in the harmonic approximation (4.55)

$$\frac{p_r^2}{2\pi} + k_F\,(r - r_e)^2 \equiv (n + \frac{1}{2})h\nu \tag{4.66}$$

$$n = 0, 1, 2, \cdots$$

Note that there is a quantum mechanical correction, equal to $h\nu/2$, to the bond energy U_0 (U_0 has a negative sign), which is called the zero-point energy. In (4.61), n is the vibrational and j the rotational quantum number. Both n and j are natural numbers.

The partition function of the diatomic molecule can be written as the product of three partition functions:

$$p.f._{\text{diat}} = p.f._{\text{trans}}\, p.f._{\text{vibr}}\, p.f._{\text{rot}}\, e^{-U_0/kT} \tag{4.67}$$

where $p.f._{\text{trans}}$ is the partition function of translational motion of the center of mass given by (4.40). Remember that the number of degrees of freedom of an atom is three, corresponding to motion in the x, y, or z direction. From the six degrees of freedom of the molecule, three have been transformed into the translational degrees of the center of the mass: one into the vibrational motion and two into the two possible rotations of the diatomic molecule. Note, however, that the separation of variables is a simplification. In fact, it is assumed that the molecule rotates as a rigid rotator with a fixed separation between the atoms equal to r_e. In reality, the atoms vibrate with respect to each other during this rotation. As a consequence, rotation and vibration are dependent strictly speaking.

The energy and entropy become a sum of translational, vibrational, or rotational contributions, because they depend logarithmically on the partition function. Because of (4.55) the vibrational partition function has been defined with respect to U_0. This can be understood from the following derivation:

$$\begin{aligned} p.f. &= \sum_i e^{-\varepsilon_i/kT} \\ &= e^{-U_0/kT} \sum_i e^{-(\varepsilon_i - U_0)/kT} \\ &= e^{-U_0/kT}\, p.f.' \end{aligned} \tag{4.68}$$

$p.f.'$ is the partition function computed with the energies of the molecules referenced to U_0. The expression for the vibrational partition function becomes

$$\begin{aligned} p.f._{\text{vibr}} &= \sum_{n=0}^{\infty} e^{-\left(n+\frac{1}{2}\right)\frac{h\nu}{kT}} \\ &= \frac{e^{-h\nu/2kT}}{1 - e^{-h\nu/kT}} \end{aligned}$$

Note that the zero of energy has been taken at the minimum of the potential well. It is customary, however, to report vibrational partition functions relative to the lowest vibrational level, which gives

$$p.f._{\text{vibr}} = \frac{1}{1 - e^{-h\nu/kT}} \tag{4.69}$$

In the classical limit this is equal to

$$p.f._{\text{vibr}} = \frac{kT}{h\nu} \quad \left(\frac{h\nu}{kT} << 1\right) \tag{4.70}$$

This expression applies for vibrations with very low frequencies (Figure 4.2). In the quantum-mechanical limit the partition function becomes

$$p.f._{\text{vibr}} = 1 \quad \left(\frac{h\nu}{kT} >> 1\right) \tag{4.71}$$

This is an important practical result as will be seen later.

The corresponding expressions for the vibrational energy and entropy (in which is included the zero point energy $h\nu/2$) become

$$\begin{aligned} \varepsilon_{\text{vibr}} &= \frac{1}{2}h\nu + \frac{h\nu}{e^{h\nu/kT} - 1} \\ &\approx kT \quad \text{classical limit} \quad \left(\frac{h\nu}{kT} << 1\right) \end{aligned} \tag{4.72}$$

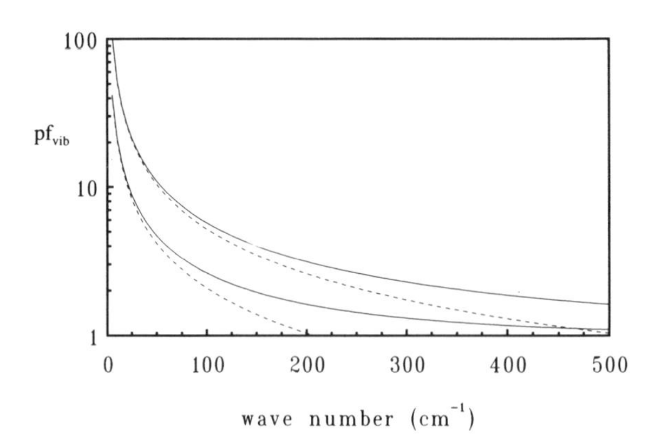

Figure 4.2. Vibrational partition function (solid) and its classical limit (dashed) as a function of the vibrational frequency, at two temperatures: 750 K (upper curves) and 300 K (lower curves).

$$s_{\text{vibr}} = k\left(\frac{h\nu}{kT}\right)\cdot\left(e^{h\nu/kT} - 1\right)^{-1} - k\ln\left(1 - e^{-h\nu/kT}\right)$$

$$\approx k\ln\left(\frac{ekT}{h\nu}\right) \quad \text{classical limit} \left(\frac{h\nu}{kT} << 1\right) \tag{4.73}$$

Result (4.72) is the same as the previously observed kT contribution of the vibrational motion to the molecular internal energy. In the quantum-mechanical limit, valid at low temperatures, there is a vibrational contribution to the energy per molecule of $\frac{1}{2}h\nu$, equal to the zero-point energy of the molecule. This is the expected result since at low temperature all molecules are predominantly in their vibrational ground state.

In order to decide at which temperature the quantum-mechanical or the classical limit applies, it is useful to define a characteristic temperature $\theta_{\text{vibr}} = h\nu/k$. The quantum-mechanical limit applies when $T << \theta_{\text{vibr}}$, the classical limit when $T >> \theta_{\text{vibr}}$. With molecules such as H_2, O_2, N_2, and CO under typical catalytic reaction conditions, this will almost always be in the quantum-mechanical situation (see Table 4.2).

The rotational partition function of a heteroatomic molecule follows from†

$$p.f._{\text{rot}} = \sum_{j=0}^{\infty} (2j+1)\exp\left(-\frac{j(j+1)h^2}{8\pi^2 IkT}\right) \qquad (a)$$

$$\approx \int_0^{\infty} d(j(j+1))\exp\left(-\frac{j(j+1)h^2}{8\pi^2 IkT}\right) \qquad (b)$$

$$= \frac{8\pi^2 IkT}{h^2} \qquad (c) \tag{4.74}$$

The factor $(2j + 1)$ appears into the summation because each rotational state j can have $(2j + 1)$ values of j_z, the projection of the rotational motion along the z axis. The final result in (4.74) is valid in the classical limit, that is at temperatures higher than $\theta_{rot} = h^2/8\pi^2 Ik$ (Table 4.2). Under this condition the difference between (4.74a) and (4.74b) is also negligible. As Table 4.2 indicates, this is the prevailing situation.

The rotational energy and entropy become

---

†The general formula for a larger molecule is $p.f._{\text{rot}} = \frac{1}{\sigma}\left(\frac{8\pi^2 kT}{h^2}\right)^{3/2}\sqrt{\pi I_A I_B I_C}$

TABLE 4.2. Typical Values of the Rotational and Vibrational Partition Functions of Diatomic Molecules at 500 K, along with Values of the Characteristic Temperatures of Rotation and Vibration, θ_{rot} and θ_{vib}[a]

Molecule	Rotational partition function	Vibrational partition function[b]	θ_{rot} (K)	θ_{vibr} (K)
H_2	2.9	1.0	85	6210
CO	180	1.0	2.8	3070
Cl_2	710	1.3	0.35	810
Br_2	2100	1.7	0.12	470

[a]Which can be used to decide whether the quantum-mechanical ($T < \theta$) or the classical limit should be applied.
[b]With respect to the lowest vibrational level.

$$\varepsilon_{rot} = kT \tag{4.75}$$

$$s_{rot} = k \ln\left(\frac{8\pi^2 eIkT}{h^2 g}\right) \tag{4.76}$$

In expression (4.76) the factor g, the symmetry factor, has been added. For a homonuclear diatomic molecule $N = 2$ and $g = 2$.

In the limit of low frequency and high temperature one has

$$\begin{aligned} \varepsilon &= \varepsilon_{trans} + \varepsilon_{vibr} + \varepsilon_{rot} + U_0 \\ &= \frac{7}{2}kT + U_0 \quad (h\nu << kT) \end{aligned} \tag{4.77}$$

This is the classical result obtained earlier. But in the low temperature limit one finds

$$\varepsilon = \frac{5}{2}kT + U_0 \quad (h\nu >> kT) \tag{4.78}$$

The corresponding specific heats are respectively $\frac{7}{2}k$ and $\frac{5}{2}k$ ($c_s = d\varepsilon/\partial T$). Historically, the fact that the classical prediction for the specific heat, $c_s = \frac{7}{2}k$, was clearly in disagreement with experimentally determined values for hydrogen formed one of the motivations for the development of quantum mechanics.

As follows from (4.72) there is only a finite contribution of the vibrational entropy in the classical limit. The contribution of the internal degrees of motion due to rotational motion is always of importance and significantly larger than that resulting from vibrational motion. A comparison of typical values of the partition function of a few representative diatomic molecules is given in Table 4.2. The reason the rotational partition function is much higher than the vibrational one, becomes evident when looking at the energy-level scheme for a typical diatomic

Figure 4.3. Vibrational and rotational levels in a typical diatomic molecule; each level contributes $\exp(-\varepsilon_i/kT)$ to the partition function. Therefore the partition function for rotation is much larger than that for vibration.

molecule such as CO, in Figure 4.3. The density of vibrational levels is low, and at low temperatures only the ground state will be populated. For each vibrational level, however, there are many (in the case of CO more than 20) rotational levels, several of which will be populated even at low temperature. As each level contributes to the partition function with a weight proportional to the probability that it is occupied, it is immediately expected that $p.f._{\text{vibr}}$ is small and $p.f._{\text{rot}}$ is larger.

Once molecules become chemisorbed, they generally lose translational and rotational freedom, as illustrated in the following example. The molecule ethylene has twelve internal degrees of freedom, as indicated in Figure 4.4a. In addition, it possesses translational and rotational freedom. Table 4.3 gives the corresponding partition functions of ethylene in the gas phase in ultrahigh vacuum (pressure approximately 10^{-10} mbar). When the molecule adsorbs on the (111) surface of rhodium or platinum, each carbon atom forms a bond with an atom in the substrate. In this di-σ coordination, the rotational and translational modes of the free ethylene have been replaced by vibrations, such as the frustrated or hindered translations (labeled 415, 355, and 70 cm^{-1} in Figure 4.4b and rotations 850, 530, and 45 cm^{-1}). Several of the vibrational frequencies of adsorbed ethylene have been measured, the others had to be estimated on measurements on the basis of coordination complexes that contain di-σ bonded ethylene. The partition functions are given in Table 4.3.

The effort invested in statistical thermodynamics so far starts to pay off, because the equilibrium constant can be predicted for a reaction from the partition functions of reactants and products. The equilibrium constant can be calculated from the analogue of the van't Hoff relation (1.9)

$$\overline{K}_{\text{eq}} = \frac{N_{AB}}{N_A N_B} \tag{4.79a}$$

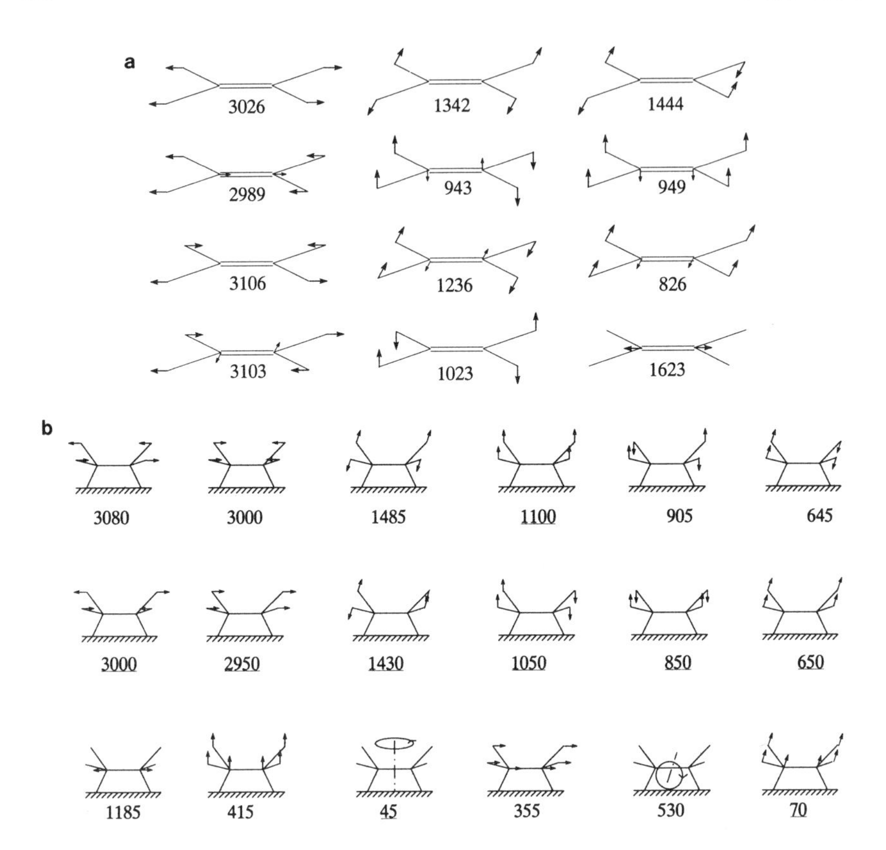

Figure 4.4. The vibration modes of gas phase ethylene (a) and di-σ bonded ethylene (b) on the (111) surface of rhodium. Frequencies in cm^{-1} are indicated, underlined values have been estimated. Gas phase data from Shimanouchi (1972), experimental data from Bent (1986) (Courtesy of R. Gelten, 1994).

TABLE 4.3. Partition Function of Gas Phase and Adsorbed Ethylene

		Gas phase ethylene				Adsorbed ethylene	
T [K]	pf_{trans}	pf_{rot}	pf_{vib}	pf'_{vib}[a]	pf_{tot}	pf_{vib}	$pf'_{vib} = pf_{tot}$
100	$3.85 \cdot 10^{18}$	$1.28 \cdot 10^{2}$	1.00	$2.67 \cdot 10^{-68}$	$1.32 \cdot 10^{-47}$	$3.33 \cdot 10^{0}$	$2.52 \cdot 10^{-71}$
300	$6.00 \cdot 10^{19}$	$6.65 \cdot 10^{2}$	1.06	$3.16 \cdot 10^{-23}$	$1.26 \cdot 10^{0}$	$3.18 \cdot 10^{1}$	$6.24 \cdot 10^{-23}$
500	$2.15 \cdot 10^{20}$	$1.43 \cdot 10^{3}$	1.44	$4.39 \cdot 10^{-14}$	$1.35 \cdot 10^{10}$	$2.50 \cdot 10^{2}$	$1.49 \cdot 10^{-12}$
700	$4.99 \cdot 10^{20}$	$2.37 \cdot 10^{3}$	2.42	$5.38 \cdot 10^{-10}$	$6.36 \cdot 10^{14}$	$1.74 \cdot 10^{3}$	$1.21 \cdot 10^{-7}$
1000	$1.22 \cdot 10^{21}$	$4.04 \cdot 10^{3}$	6.30	$1.10 \cdot 10^{-6}$	$5.42 \cdot 10^{18}$	$2.47 \cdot 10^{4}$	$1.91 \cdot 10^{-3}$

[a]This is the vibrational partition function with correction for the zero-point vibrational energy. Rotational partition function has been calculated according to the formula

$$pf_{rot} = \frac{1}{\sigma}\left(\frac{8\pi^2 kT}{h^2}\right)^{3/2} \sqrt{\pi I_A I_B I_C}$$

in which $\sigma = 4$ and the moments of inertia are 5.75×10^{-47}, 28.09×10^{-47}, and 33.84×10^{-47} kgm^2, according to Gallaway and Barker (1942).

$$= e^{-\sum_i \nu_i \bar{\mu}_{0i}/kT} \tag{4.79b}$$

$$= e^{-\Delta G_0/RT} \tag{4.79c}$$

Because it depends on numbers, $\overline{K}$ is dimensionless, whereas K_{eq} defined in (4.78b) depends on concentrations and has the dimension of V^{-1}

$$\overline{K}_{eq} = V\,K_{eq} \tag{4.80a}$$

$$\text{with}\;\; K_{eq} = \frac{[N_{AB}]}{[N_A][N_B]} \tag{4.80b}$$

Using the relation between the chemical potential and the partition function, (4.35) gives

$$\overline{K}_{eq} = \frac{\prod_i (p.f.)^{\nu_i}_{i,\text{product}}}{\prod_j (p.f.)^{-\nu_j}_{j,\text{reactant}}} \tag{4.81}$$

where it should be remembered that the stoichiometric constants are positive for products and negative for reactants.

Although expression (4.81) may be a bit cryptic at first sight, it provides insight in chemical equilibria right away, as the following example illustrates.

Consider the equilibrium between two isomers

$$\mathrm{A} \underset{k_1^f}{\overset{k_1^b}{\leftrightarrows}} \mathrm{B} \tag{4.82}$$

and suppose that the ground state energy of A is lower than that of B, while the energy levels of B are closer together (Figure 4.5). This would be the case if the chemical bonds in A are stronger than in B, as a result of which vibrational frequencies of B are smaller than in A and rotational moments are larger, because the atoms in B are at a larger distance from each other than in A. Note that the translational energies of A and B are equal. According to (4.81), the equilibrium constant for this process is

$$\overline{K}_{eq} = \frac{p.f._{\mathrm{B}}}{p.f._{\mathrm{A}}} = \frac{\sum_i e^{-\varepsilon_i^{\mathrm{B}}/kT}}{\sum_i e^{-\varepsilon_i^{\mathrm{A}}/kT}} \tag{4.83}$$

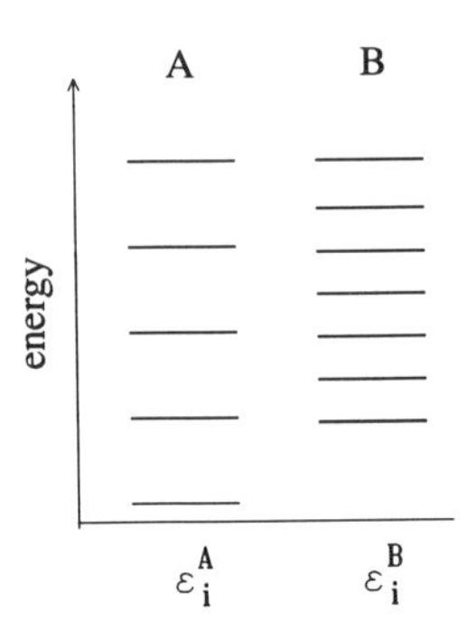

Figure 4.5. The energy states of two isomers, A and B.

The temperature will determine which of the two isomers is dominant. The first term in $p.f._A$ is the largest of all, which causes $\overline{K}_{eq}$ to be smaller than one at low temperature, in favor of isomer A. However, the number of terms in $p.f._B$ is larger than in $p.f._A$, so that $\overline{K}_{eq}$ may become larger than one at higher temperatures. In general, the equilibrium state that has low lying and closely spaced levels will dominate. The former is an energy effect, the latter an entropy effect. Thus, if the equilibrium concerns a competition of relatively stable reactants and less stable products of high entropy, the reactants are favored at low temperature and the products at high temperature. At intermediate temperatures, the equilibrium composition is determined by a compromise between low energy and high entropy to give the system the minimum free energy $E - TS$.

Consider the equilibrium constant of the homonuclear association–dissociation reaction

$$A + A \underset{k_1^f}{\overset{k_1^b}{\leftrightarrows}} A_2 \tag{4.84}$$

The equilibrium constant of this reaction becomes

$$\overline{K}_{eq} = \frac{p.f._{A_2}}{(p.f.)_A^2/2} \tag{4.85}$$

in which the factor 1/2 is included to prevent double counting. Inserting the partition function gives

$$\overline{K}_{eq} = \frac{V\left(\dfrac{2\pi M_{cm}kT}{h^2}\right)^{3/2}\left(\dfrac{e^{-h\nu/2kT}}{1-e^{-h\nu/kT}}\right)\left(\dfrac{8\pi^2 IkT}{2h^2}\right)}{\dfrac{1}{2}V^2\left(\dfrac{2\pi mkT}{h^2}\right)^3} e^{-U_0/kT} \tag{4.86}$$

Approximating (4.86) in the classical ($p.f._{\mathrm{vibr}} \approx kT/h\nu$) and in the quantum-mechanical limit ($p.f._{\mathrm{vibr}} \approx 1$) reveals

$$\overline{K}_{\mathrm{eq}}^{qm} = \frac{1}{V}\left(\frac{2\pi h^2}{kT\,M_{cm}^3}\right)^{1/2} \mu r_e^2\, e^{-(U_0+\frac{1}{2}h\nu)/kT} \tag{4.87}$$

and

$$\overline{K}_{\mathrm{eq}}^{\mathrm{class}} = \frac{kT}{h\nu} \times \overline{K}_{\mathrm{eq}}^{qm} \tag{4.88}$$

Again, the combination of temperature and vibrational frequency determines which expression for the equilibrium constant is appropriate. The values of the characteristic temperatures in Table 4.2 can be used for this purpose. H_2 and CO need the quantum-mechanical limit; the classically predicted equilibrium constant would be far too low at prevailing catalytic conditions ($<<$ 1000 K). For heavy molecules, such as Br_2 at temperatures in the order of 500 K, however, use the complete expression (4.86).

Equation (4.83) shows the expected result that molecule formation is aided by a strong bond energy. The bond energy equals the potential energy minimum corrected for the zero point vibration energy. The larger the rotational moment the higher the equilibrium constant and the more association becomes likely. As shown from the classical limit expression, a low vibration frequency encourages molecule formation. This could favor molecule formation of heavy atoms versus light atoms. From expression (4.83) the reaction enthalpy can be calculated from

$$\Delta\overline{H} = kT^2 \frac{\partial}{\partial T} \ln K_{\mathrm{eq}} \tag{4.89}$$

The resulting expressions in the classical and quantum-mechanical limit become respectively

$$\Delta\overline{H} = U_0 - \frac{1}{2}kT \quad \text{(classical limit)} \tag{4.90a}$$

$$\Delta\overline{H} = U_0 + \frac{1}{2}h\nu \quad \text{(quantum-mechanical limit)} \tag{4.90b}$$

The classical enthalpy of the reaction contains a term proportional to kT, which derives from the loss of one degree of freedom when two atoms associate.

All relations found so far for the partition functions are summarized in Table 4.1.

4.3. MICROSCOPIC EXPRESSIONS FOR THE RATE CONSTANT

4.3.1. The Rate Expression

The rate constant is given by the number of collisions between the molecules multiplied by the probability of reaction. The number of collisions follows from the average velocity of the molecules and their effective cross section for collision. The energy distribution over the different degrees of freedom of the molecules is equal to the Boltzmann distribution as long as the number of collisions between the molecules is large compared to their probability of reaction.

Consider an elementary reaction of the type

$$A + B \underset{k^f}{\overset{k^b}{\leftrightarrows}} C + D \tag{4.91}$$

An expression will be derived for the rate constant k^f. Clearly, once k^f is known, k^b follows from the relation between equilibrium constant and rate constants (see (2.3)). The rate constant can be computed from the cross section σ_r or a collision of molecule A with B, and the probability that the collision leads to a reaction.

In general, A and B are molecules which consist of two or more atoms. Each molecule has vibrational and rotational degrees of freedom, denoted with quantum numbers i and j. In addition, the centers of mass of A and B have velocities u_A and u_B, and a relative velocity between A and B can be defined as

$$\vec{u} = \vec{u}_A - \vec{u}_B \tag{4.92}$$

When A and B react, they necessarily collide. The energy involved in the collision will be redistributed over the translational (kinetic), vibrational and rotational energy of the product molecules, in a way that total energy and momentum are conserved.

The cross section can be determined experimentally from molecular beam experiments and can, also in principle, be computed. However, rigorous calculations are only possible for very simple systems (e.g., $D + H_2$). Approximations will have to be used to derive rate constants for all other chemically interesting systems. This is the subject of this section.

Not all collisions between A and B will be reactive. In fact, most collisions lead to energy transfer rather than product formation. In a symbolic notation

$$A(i, u_A) + B(j, u_B) \Rightarrow A(m, u'_A) + B(l, u'_B) \tag{4.93}$$

where m and l indicate the quantum states and u' the velocity after collision. Figure 4.5 illustrates this with a simple case. Such processes can be specified by an energy transfer cross-section.

The differential cross section $\sigma(ml/ij;\ \Omega;\ u)$ is defined by the number of particles that scatter from a state characterized by i, j and a particular solid angle Ω, into states m and l, when they approach with relative velocity u. The indices m, l, i, and j are, for instance, vibrational or rotational quantum numbers.

The number of product molecules A per unit of volume that come out of the collision in state m and scattered with solid angle Ω is given by $n_{A(m)}(\Omega)$ and follows from

$$\frac{dn_{\mathrm{A}(m)}(\Omega)}{dt} = \sigma(ml \mid ij;\ \Omega,u)\ un_{\mathrm{A}(i)} f_{\mathrm{A}(i)}\ (u_{\mathrm{A}})\ n_{\mathrm{B}(j)} f_{\mathrm{B}(j)}\ (u_{\mathrm{B}})\ d\Omega\ du_{\mathrm{A}}\ du_{\mathrm{B}} \tag{4.94}$$

in which $n_{\mathrm{A}(i)}$ and $n_{\mathrm{B}(i)}$ are the number densities of particles A and B; $f_{\mathrm{A}(i)}(u\mathrm{A})$ and $f_{\mathrm{B}(i)}(u\mathrm{B})$ the velocity distribution functions of state i and j as well as velocities u_{A} and u_{B}; u is the relative velocity of the particles; $d\Omega$ equals d cos $\theta d\varphi$ (polar coordinates); and θ and φ define the direction in which the particle is scattered. In a chemical reaction the interest is not usually in the differential cross section but in the total cross section

$$\sigma(ml \mid ij;\ u) = \int \sigma\ (ml \mid ij;\ \Omega u)\ d\Omega \tag{4.95}$$

If the collision can be described according to classical mechanics, the total cross section can be rewritten as

$$\sigma(ml \mid ij;\ u) = 2\pi \int_0^\infty P\ (ml \mid ij;\ u\, b) b\ db \tag{4.96}$$

where $P(ml \mid ij;u,\ b)$ is the probability of reaction. The flux of the relative motion is given by the impact parameter b. The impact parameter is defined as the closest distance of approach of the centers of two molecules (on the assumption of no interaction). This is illustrated in Figure 4.6. Particles are considered as hard spheres that move with an average relative velocity u.

All distances of approach have to be integrated over. Therefore integrate over the area of possible impact parameters b. If the impact parameter b is larger than $d = (d_{\mathrm{A}} + d_{\mathrm{B}})/2$, d_{A} and d_{B} being the diameters of particles A and B, no collision occurs. Per second the colliding molecule moves through a cylinder of cross sectional area πd^2 and length u. Every molecule that overlaps at least partially with this cylinder will be struck. So only for impact parameter values smaller than $b_{\max}$ will there be a contribution to the cross section

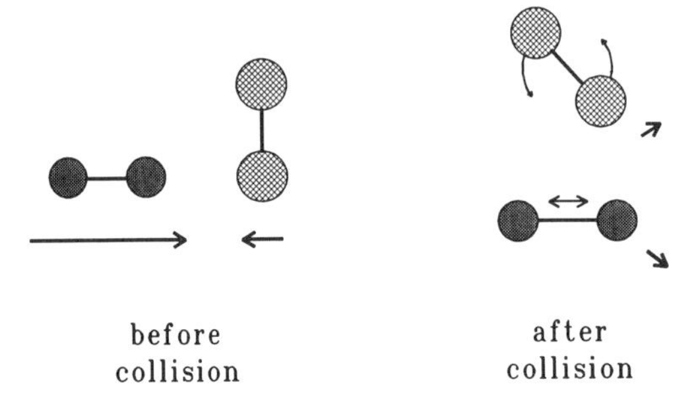

Figure 4.6. Energy transfer in a collision between two diatomic molecules. For simplicity we assume that before the collision the molecules are in their vibrational and rotational ground states and possess translational energy only. After collision the molecules are vibrationally and rotationally excited and move at lower velocity.

$$\sigma(ml \mid ij;u) = 2\pi \int_0^{b_{\max}} P(ml \mid ij;\, u\, b) b\, db \tag{4.97}$$

Equation (4.97) will be returned to later to evaluate σ. The expression for the rate of production of molecules C becomes

$$\frac{dn_C}{dt} = -\frac{dn_A}{dt} = \sum_{ijlm} \int \int u\sigma(ml \mid ij;\, u) f_{A(i)}(u_A)\, n_A(i)\, f_{B(j)}(u_B)\, n_B(j)\, du_A\, du_b \tag{4.98}$$

If the number density of A(i) is written as

$$n_{A(i)} = n_A\, x_{A(i)} \tag{4.99}$$

with $x_{A(i)}$ the distribution of states i in molecule A, expression (4.98) can be rewritten as a macroscopic rate equation

$$\frac{dn_C}{dt} = k_\sigma\, n_A\, n_B \tag{4.100}$$

with

$$k_\sigma = \sum_{ijlm} x_{A(i)}\, x_{B(j)} \int \int u\sigma(ml \mid ij;\, u) f_{A(i)}(u_A)\, f_{B(j)}(u_B) d\vec{u}_A\, d\vec{u}_B \tag{4.101}$$

When the number of non-reactive collisions between the molecules is large compared to the number of reactive collisions, the molecules are in thermal equilibrium with their surroundings and therefore the internal state distribution functions $x_{A(i)}$ and $x_{B(j)}$ can be taken as Boltzmann distributions. The same holds for the velocity distribution functions.

The important principle of different time scales is met here. It may be assumed that the energy is distributed according to the Boltzmann distribution, when the overall rate of the reaction is small compared to the number of collisions that are unreactive, i.e., only exchange energy.

In practice, reactions are carried out in a liquid, a dense gas, or on a surface. In such cases, collisions between the reacting molecules and the surrounding molecules are important as well. Such three particle interactions can help to maintain equilibrium in the internal degrees of freedom of the reacting complex A–B. As will be revealed, the rate constants derived for the case of two-particle interactions remain often applicable for reactions in media. It is somewhat ironic that the presence of three-particle interactions can be a necessary condition for the validity of two-particle rate expressions, especially when equilibration of internal degrees of freedom has been assumed. This will be discussed in more detail in Chapter 5.

When the molecules are equilibrated, the energies of their center mass motion is given by the Boltzmann expression

$$f_{A(i)} = \frac{e^{-P_A^2/2m_A kT}}{p.f._A} \tag{4.102}$$

and is independent of the state of molecule i. According to (4.101) we have to evaluate

$$d\,\vec{u}_A\; d\,\vec{u}_B\; u\,\sigma\, f_A(\vec{u}_A)\, f_B\,(\vec{u}_B) \tag{4.103}$$

This expression can be written as the product of the partition functions of the relative motion of molecules A and B and that of their center mass motion as discussed in (4.52). Expression (4.103) can be substituted by

$$d\,\vec{u}\,d\,\vec{u}_{cm}\; u\,\sigma\, f(u_{cm})\, f(u) \tag{4.104}$$

Performing the integration gives as final expression for k_σ

$$k_\sigma = \sum_{ij,lm} x_{A(i)}\, x_{B(j)} \int u\sigma(ml\,|\,ij;\, u)\, f(u)d\,\vec{u} \tag{4.105}$$

This is a general and not very practical form of the reaction-rate constant. Two important special cases are treated in the next sections.

4.3.2. Collision Theory of Reaction Rates

In collision theory molecules are considered as hard spheres which possess only translational energy and have no internal degrees of *freedom. Reaction occurs if the translational energy involved in the collision is higher than a certain energy barrier, E_b. The reaction-rate constant is the product of the collision frequency and a Boltzmann factor,* $\exp(-E_b/kT)$ *which expresses that only molecules with translational energy lower than E_b react. In general, hard-sphere collision theory provides an upper limit of the reaction rate.*

Consider a reaction between molecules A and B. The rate expression for molecules that are essentially billiard balls, or hard spheres, will be discussed first. Later studied will be features due to the presence of internal degrees of freedom, which are absent in hard spheres. The description of reaction rates in terms of the kinetic theory of collisions was given by Max Trautz in 1916 and William Lewis in 1918. The rate for collisions between hard spheres is

$$r_{\text{coll}} = k_\sigma N_{\text{A}} N_{\text{B}}$$

with

$$k_\sigma = \pi d^2 \frac{1}{g_{\text{AB}}} \frac{\int_0^\infty \int_{-1}^{+1} \int_0^{2\pi} u\, e^{-\mu u^2/2kT} u^2 \, du \, d\cos\Theta \, d\varphi}{\int_0^\infty \int_{-1}^{+1} \int_0^{2\pi} e^{-\mu u^2/2kT} u^2 \, du \, d\cos\Theta \, d\varphi}$$

$$= \frac{\pi d^2 \bar{u}}{g_{\text{AB}}} \tag{4.106}$$

where N_{A} and N_{B} represent the number of molecules in the reacting volume, and $d = \frac{1}{2}(d_{\text{A}} + d_{\text{B}})$, the average diameter of A and B. The symmetry factor g_{AB} has been introduced to prevent double counting in case the reaction occurs between molecules of the same type. Thus, g equals 2 for reactions between A and A, and 1 for reactions between A and B.

Expression (4.106) follows from (4.105) by a) omitting the summation over i, j, l, and m and putting x to 1 because the particles have no internal degrees of freedom; b) using the expression for the Boltzman distribution function of the relative motion of particles A and B; and c) use of expression (4.97) to evaluate the cross section. For a collision of hard spheres:

$$P(b) = 1 \quad \text{if} \quad b \le \frac{1}{2}(d_{\text{A}} + d_{\text{B}}) \tag{4.107a}$$

$$P(b) = 0 \quad \text{if} \quad b > \frac{1}{2}(d_A + d_B) \tag{4.107b}$$

where b is the impact parameter (Figure 4.6).

The average velocity between the particles can readily be evaluated and is

$$\bar{u} = \left(\frac{8kT}{\pi\mu}\right)^{1/2} \tag{4.108}$$

A typical order of magnitude is 5×10^4 cm/sec. It follows that the number of hard sphere collisions in a gas equals

$$r_{\text{coll}} = \frac{\pi d^2}{g_{AB}} \left(\frac{8kT}{\pi\mu}\right)^{1/2} N_A N_B \tag{4.109}$$

Thus, the number of collisions between A and B in a gas of 1 bar pressure is in the order of 10^{28} per cm^3 and per second. Note that through the term $N_A N_B$ the rate r_{coll} depends quadratically and thus strongly on the pressure, but only moderately on temperature. The actual rate of reaction between A and B is many orders of magnitude smaller than r_c, so that the vast majority of collisions is unreactive.

Since catalysis involves molecules adsorbed on a surface, it is also useful to consider the expression for the number of hard sphere collisions with a surface (Figure 4.7). Then only the velocity component of a particle into the direction of a surface is of interest, and only values of the angle θ smaller than $\pi/2$ contribute to a collision and the relative velocity $\bar{u}_s$ for surface collisions becomes

$$\bar{u}_s = \frac{\int_0^1 d\cos\Theta \int_0^{2\pi} d\varphi \int_0^\infty du\, u^2\, u \cos\Theta\, e^{-mu^2/2kT}}{\int_{-1}^1 d\cos\Theta \int_0^{2\pi} d\varphi \int_0^\infty du\, u^2\, e^{-mu^2/2kT}} \tag{4.110}$$

$$= \frac{1}{4}\left(\frac{8kT}{\pi m}\right)^{1/2}$$

Note that expression (4.110) does not contain the reduced mass μ, of the colliding particle but the mass of a freely moving particle. The surface mass can be considered infinite compared to colliding molecule mass. The average collision velocity is reduced by a factor 4 compared to that for gas phase atoms.

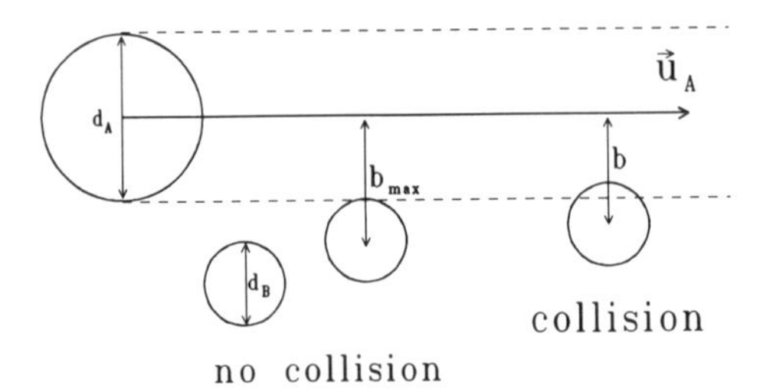

Figure 4.7. The impact parameter: collision occurs if the impact parameter b is smaller than $(d_A + d_B)/2$.

For hard sphere collisions between molecules and a surface, an expression very similar to (4.106) can be deduced as follows. The number of surface collisions per unit of time is

$$-\frac{dN_g}{dt} = S \cdot \bar{u}_s \cdot \frac{N_g}{V} \tag{4.111}$$

where S is the surface area and N_g the number of molecules in a volume V above the surface. Let the diameter of the colliding molecule be d. The maximum number of particles N_{max} that can adsorb, is equal to

$$N_{max} = \frac{S}{\frac{1}{4}\pi d^2} \tag{4.112}$$

and the surface coverage $\theta = N_{ads}/N_{max}$.

The rate of coverage increase is then given by

$$\frac{d\theta}{dt} = \frac{1}{4}\pi d^2 \bar{u}(1-\theta) \cdot \frac{N_g}{V} \tag{4.113}$$

where the term $(1-\theta)$ accounts for the surface fraction that is empty. The expression for the rate constant of adsorption for hard sphere collisions becomes

$$k_{ads}^{coll} = \frac{1}{4}\pi d^2 \bar{u} \tag{4.114}$$

which is the equivalent of (4.106) for adsorption on a surface.

So far, the hard sphere molecules can only collide and scatter. Consider the expression for k_σ if collisions occur between molecules that not only experience a repulsive, but also an attractive potential. At infinite distance the kinetic energy is given by

$$E = \frac{1}{2}\mu u^2 \tag{4.115}$$

where u is the relative velocity between the particles. E in (4.115) is the initial energy in the problem to be discussed. This amount of energy is conserved in the collision. When the two particles interact, the energy becomes distributed over several terms

$$E = \frac{1}{2}\frac{p_r^2}{\mu} + \frac{1}{2}\frac{p_a^2}{\mu r^2} + V(r) \tag{4.116}$$

The first term is the relative kinetic energy, the second the centrifugal energy, due to the finite angular moment (see (4.63)) and the last term is the potential energy $V(r)$. Conservation of the angular momentum gives

$$p_a = \mu u b \tag{4.117}$$

b is the impact parameter as defined in Section 4.3.1.

Expression (4.117), together with conservation of energy gives

$$\frac{1}{2}\mu u^2 = \frac{1}{2}\mu \dot{r}^2 + \frac{1}{2}\mu u^2\left(\frac{b}{r}\right)^2 + V(r) \tag{4.118}$$

Equation (4.118) can be used to find $r_{\min}$, the distance of closest approach, at which the relative particle velocity changes sign. This point is given by the value of r for which the kinetic energy becomes zero

$$\frac{1}{2}\mu u^2 = \frac{1}{2}\mu u^2\left(\frac{b}{r_{\min}}\right)^2 + V(r_{\min}) \tag{4.119}$$

revealing

$$\left(\frac{b}{r_{\min}}\right)^2 = 1 - \frac{V_{\min}}{\frac{1}{2}\mu u^2} \tag{4.120}$$

so that reaction only occurs when

$$\begin{aligned} b^2 &= r_{\min}^2\left(1 - \frac{(V(r_{\min})}{E}\right) \\ &\le r_{\min}^2 \end{aligned} \tag{4.121}$$

The maximum value of the impact parameter for which reaction occurs depends on the value of $V(r_{min})$.

Usually chemical reactions require some kind of energy activation in order to proceed. In a chemical reaction, bonds have to be broken before new ones are formed. Consider HCl formation from H_2 and Cl_2 as an example

$$H_2 + Cl_2 \rightarrow 2\,HCl \tag{4.122}$$

The reaction is exothermic, but when it proceeds via collision of H_2 and Cl_2 in the gas phase it will only do so at high temperatures, because the bonds of Cl_2 (240 kJ/mol) and H_2 (430 kJ/mol) have to be broken. Usually the energy required is much less than the full bond energies, formation of intermediate complexes between the reacting molecules will create low energy reaction paths.

In the case of HCl formation, the reaction is initiated by chlorine dissociation. The full bond energy of 240 kJ/mol has to be overcome, which is achieved by the adsorption of light

$$Cl_2 \underset{h\nu}{\rightarrow} 2Cl \tag{4.123}$$

Once a few molecules have been initiated, the reaction proceeds as a chain process

$$Cl + H_2 \rightarrow HCl + H \tag{4.124a}$$

$$H + Cl_2 \rightarrow HCl + Cl \tag{4.124b}$$

etc.

These reaction steps proceed via reaction intermediates such as

$$Cl \cdots H \cdots H$$

Formation of the new bond assists in breaking the old bond by lowering the overall energy that is required.

Figure 4.8 illustrates the reactions in a potential energy diagram. The distance Cl–H_2 has been taken as the reaction coordinate. The relative velocity $\vec{u}$ equals $\vec{u}_{H_2} - \vec{u}_{Cl}$, and the reduced mass $\mu = M_{H_2} \cdot M_{Cl}/(M_{H_2} + M_{Cl})$. A reaction occurs if the kinetic energy $(\frac{1}{2}\mu u^2)$ involved in the collision between Cl and H_2 is larger than the energy of the barrier, E_b. The latter represents the energy needed to bring the reactants Cl and H_2 together in the complex $Cl \cdot\cdot H \cdot\cdot H$, where simultaneously with the breaking of the H–H bond, the new Cl–H bond forms.

In a catalytic reaction, surface-complex formation plays a role similar to reaction-intermediate formation. For instance, on Cu the reaction can be thought to proceed as follows

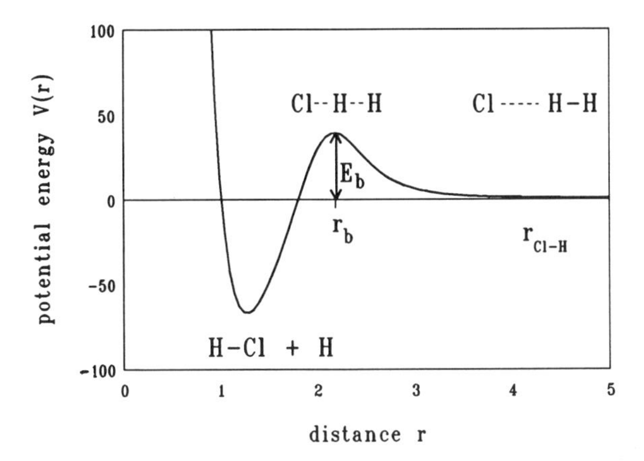

Figure 4.8. Potential energy diagram for the reaction between Cl and H_2 in the collision theory. The energy in the collision between Cl and H_2, $\frac{1}{2}\mu u^2$, must be at least E_b for reaction to occur.

$$Cl_2 + 2Cu \rightarrow 2CuCl \tag{4.125a}$$

$$2CuCl + H_2 \rightarrow 2HCl + 2Cu \tag{4.125b}$$

Instead of using light, the heat of formation of CuCl provides the driving force for breaking the chlorine bond. This book will later cover how transition-state theory enables the calculation of reaction-rate constants from information on the intermediate complex formed. However, the simplest expression for the rate constant follows from the assumption that collisions are essentially hard sphere, with a total cross section given by πd^2. However, as long as the relative energy is below a critical value E_b no reaction can take place. Reaction only occurs if $E \geq E_b$, (Figure 4.8). The expression for the effective cross section then becomes (see 4.96, 4.121)

$$\begin{aligned} \sigma &= \pi b_{\max}^2 && (a) \\ &= \pi d^2 \left(1 - \frac{E_b}{E}\right) \quad E \geq E_b;\ P = 1 && (b) \\ &= 0 \quad E < E_b;\ P = 0 && (c) \end{aligned} \tag{4.126}$$

in which P stands for the probability that the reaction occurs. Substitution of (4.126) into (4.105) together with the assumptions (a) and (b) made in the derivation of (4.106) gives

$$k_\sigma(T) = \frac{1}{g_{\mathrm{AB}}} \frac{8\pi\mu}{(2\pi\mu kT)^{3/2}} \int_{E_b}^{\infty} \pi d^2 \left(1 - \frac{E_a}{E}\right) E\, e^{-E/kT}\, dE$$

$$= \frac{1}{g_{\mathrm{AB}}} \pi d^2 \left(\frac{8kT}{\pi\mu}\right)^{1/2} e^{-E_b/kT} \tag{4.127}$$

The term before the integration sign of (4.127) stems from the transformation

$$\frac{1}{\mu} p^3\, dp = 2\mu E\, dE \tag{4.128}$$

Expression (4.127) has a clear interpretation if rewritten in the form of the rate for the reaction between A and B

$$r = \frac{1}{g_{\mathrm{AB}}} \pi d^2 \left(\frac{8kT}{\pi\mu}\right)^{1/2} e^{-E_b/kT} N_{\mathrm{A}}\, N_{\mathrm{B}} \tag{4.129}$$

One recognizes the rate of hard-sphere collision between A and B, multiplied by a Boltzmann factor to express that only molecules which collide with energy larger than E_b react. As stated, the collision rate is in the order of 10^{28} for a gas at 1 bar pressure. The Boltzmann factor represents the fraction of collisions that leads to reaction. This fraction usually varies between 10^{-10} and 10^{-20}, depending on the height of the energy barrier.

Expressions (4.127) and (4.129) have the same form as the Arrhenius equation

$$k_{\mathrm{Arrhenius}} = \nu_{\mathrm{eff}}\, e^{-E_{\mathrm{act}}/kT} \tag{4.130}$$

Note, however, that the entire temperature dependence of the rate reaction in its Arrhenius form is in the exponential term, whereas (4.129) contains an additional temperature dependence as a result of the collision frequency. It is therefore not correct to take the barrier energy E_b of (4.129) equal to the experimentally determined activation energy E_{act}. In general, E_{act} can be computed from (4.127) using its definition

$$E_{\mathrm{act}} = \Delta H_a = -k \frac{\partial}{\partial\left(\frac{1}{T}\right)} \ln k_{\mathrm{Arrhenius}} \tag{4.131a}$$

$$= kT^2 \frac{\partial}{\partial T} \ln k_{\mathrm{Arrhenius}} \tag{4.131b}$$

Note the similarity of (4.36) to (4.131*b*). The rate constant replaces the partition function in (4.36). Applying relation (4.131) to $k_\sigma(T)$ of (4.127) gives for the activation enthalpy

$$E_{act} = \Delta H^{hardsphere} = E_b + \frac{1}{2}kT \tag{4.132}$$

When energies are expressed per gmol, the Boltzmann constant is replaced by the gas constant R in expressions relating to energy. The effective frequency ν_{eff} in the Arrhenius rate equation will be discussed later.

Thus far we have discussed reactions as occurring between hard reactive spheres. The rate constant is found to be the product of a cross section of the colliding molecules, a mean molecular velocity, and a Boltzmann factor. It is the latter that accounts predominantly for the temperature dependence of the reaction rates, as the mean molecular velocity depends on temperature only as $T^{1/2}$. The only type of energy involved in the collisions is translational energy, because hard spheres possess no internal degrees of freedom which can exchange energy. As a consequence, the theory treats reaction rates in terms of energy barriers and not in terms of *free* energy barriers. This shortcoming is immediately apparent if the equilibrium constant from the collision theory expressions for the forward and reverse rate constants of a reaction is calculated. This gives a relation between K^{eq} and the heat of reaction, but not between K^{eq} and the change in free energy, as van't Hoffs equations require. Finally, the vast majority of collisions does *not* lead to a reaction because their translational energy is too low.

4.3.3. Transition-State or Activated Complex Theory

Most reactions occur only if the bonds in a molecule are activated. In other words, most reactions have an activation barrier. When the activation energy is high compared to the thermal energy, kT, the probability of activation of a molecule is low. Many collisions are needed to provide enough energy. Once a molecule is activated, the passage over the barrier is fast. According to transition state theory the rate constant of a reaction is (kT/h) $K^{\#}$, where kT/h represents the rate of a molecule passing the activation barrier, and $K^{\#}$ the equilibrium constant of the activated complex.

Transition state, sometimes termed activated complex, theory yields kinetic expressions that are applicable over a wide range of reaction conditions and has been extensively used in chemistry. The theory was developed by Eyring, and independently by M.G. Evans and Michael Polanyi, around 1935. Before deriving an expression for the rate constant, we give a more qualitative description of the transition-state theory (based on Laidler, 1987).

The reaction is thought to proceed through an activated complex, the transition state, located at an energy barrier separating reactants and products. It can be visualized by the travel over a potential energy surface, such as a mountain landscape where the barrier lies at the saddle point, the mountain pass or *col*. The event is described by one degree of freedom (e.g., a vibration in case of a dissociation reaction) called the reaction coordinate, $q^{\#}$. The rate expression of the transition-state theory rests on the following assumptions:

1. Reaction systems pass the barrier in only one direction. Once over the col they cannot turn back.
2. The energy distribution of the reactant molecules is given by the Boltzmann distribution. Those activated complexes that are becoming products are essentially in equilibrium with the reactants, except with respect to the reaction coordinate. This assumption implies that the concentration of transition complexes may be expressed in terms of partition functions and an energy difference, in accordance with (4.81) and (4.68).
3. The passage over the barrier is the motion of only one degree of freedom, the reaction coordinate, which is independent of all other motions of the activated complex.
4. The passage over the barrier is treated as a classical event; all quantum effects are ignored. This assumption makes the conventional transition-state theory a "hybrid" formalism, as also quantum-mechanical expressions appear in the rate expression, through the partition functions involved in assumption 2.

These are the main ingredients of the transition-state theory. They are summarized in Figure 4.9.

In order to derive the expression for the rate constant of reacting molecules with internal degrees of freedom, expression (4.105) is reconsidered. For the reaction between A and B the rate constant can be rewritten as (Johnston, 1966)

$$k_{\sigma} = \frac{1}{p.f._{\mathrm{AB}}^{\mathrm{rel}}} \frac{1}{h^{3N-3}} \underset{6N-6}{\int \int} ds\; v(s) \times P_b(s)\; e^{-\varepsilon(s)/kT} \tag{4.133}$$

where $p.f._{\mathrm{AB}}^{\mathrm{rel}}$ is the partition function of the relative motion of the center of mass of the reacting molecules and N is the number of atoms in the colliding complex.

The integral of (4.133) is the $6N-6$ dimensional integral over the phase space determined by the atoms of the two molecules minus the center of mass motion.

$$\begin{aligned} ds &= ds_1 \;\ldots\ldots\; ds_{3N-3} \\ &= dp_1\, dq_1 \;\ldots\ldots\; dp_{3N-3}\, dq_{3N-3} \end{aligned} \tag{4.134}$$

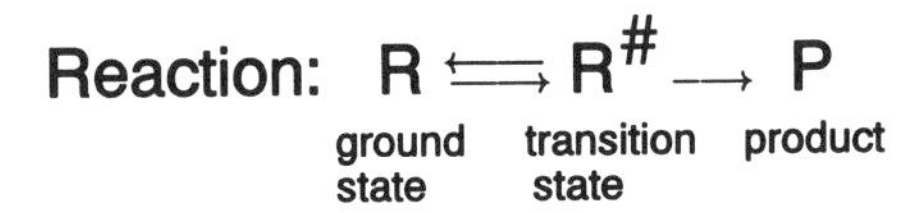

Assumptions:

Passage over the barrier only in forward direction

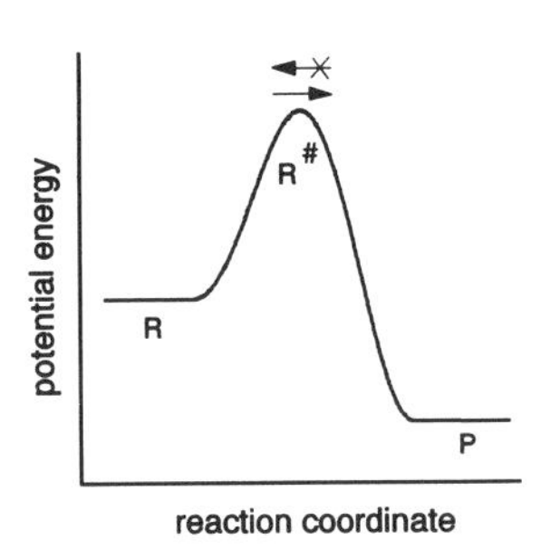

Equilibrium between ground state and transition state for all degrees of freedom, except for the reaction coordinate $q^{\#}$

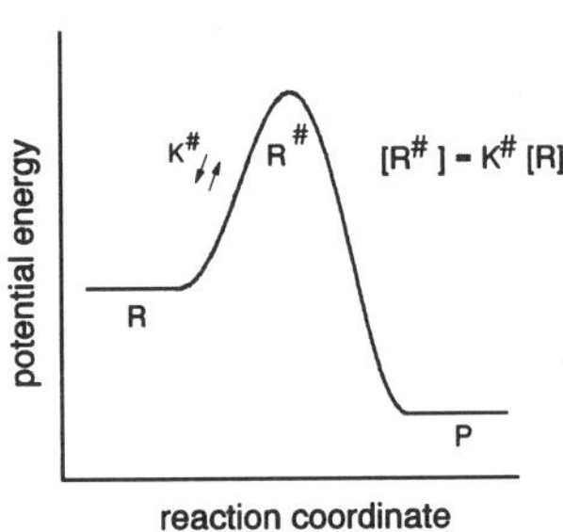

Passage over the barrier is a classical event, described by the motion of one reaction coordinate only

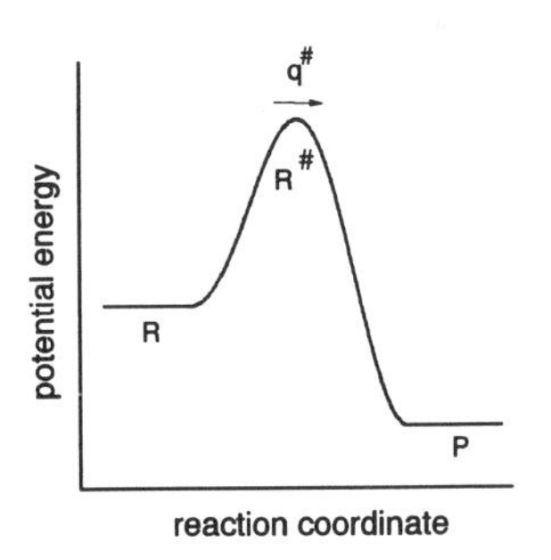

Then: $R \overset{K^{\#}}{\rightleftharpoons} R \overset{kT/h}{\longrightarrow} P$

and $\dfrac{d\,[P]}{dt} = \dfrac{kT}{h} K^{\#} [R]$

Figure 4.9. The basic assumptions of the transition-state theory.

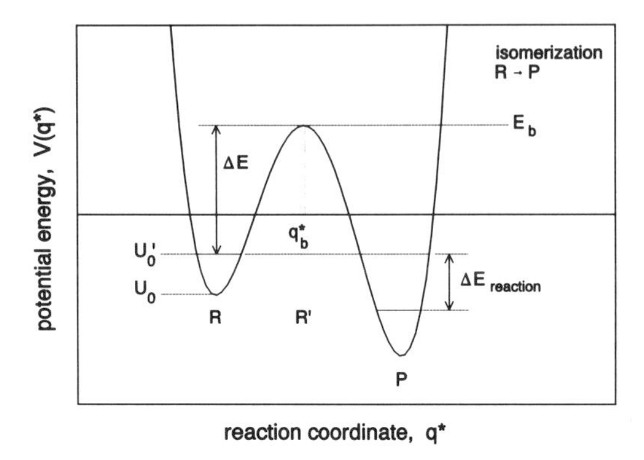

Figure 4.10. Reaction energy diagram as a function of reaction coordinate q^* for an isomerization reaction. U_0 is the potential energy minimum of R and U'_0 the potential energy minimum corrected for the zero point vibration energy.

$\nu(s)$ is the collision frequency of the molecules and $P_b(s)$ the probability of reaction.

The transition-state expression for the reaction rate is derived using the following approximating assumptions

Assumption I: *The frequency of crossing the surface of separation of products from reactants* $\Delta E = E_b - U'_0$ *depends on only one degree of freedom, with coordinate* q^* *and momentum* p^*. *This degree of freedom is called reaction coordinate.*

For an isomerization reaction, a representation of the potential energy dependence of the reaction on reaction coordinate q^* is given in Figure 4.10. The activation energy is given by the difference in energy of E_b, the energy at q_b^*, and the energy of the reactant U'_0. The overall reaction energy $\Delta E_{\text{reaction}}$ depends on the energy difference of product and reactant. Assumption I does not imply that there is no change in the configurations determined by the other degrees of freedom. As will be seen the configuration change is often such that it optimizes the products $\nu(s) \times P_b(s)$.

A consequence of assumption I is that the collision complex does not undergo collision with a third body during its motion over the barrier. This will be analyzed in more detail in Chapter 5. This assumption implies that the characteristic time τ_b of motion over the barrier is short compared to the collision time of non-reactive collisions. The latter has to be small compared with the overall reaction time τ_r:

$$\tau_r = k_\sigma^{-1} \tag{4.135}$$

so that the time scales for transition-state theory to be valid become

$$\tau_b < \tau_{\text{coll}} < \tau_r \tag{4.136}$$

Relation (4.136) contains the previously mentioned basic postulate concerning the energy exchange time scale between the molecules during reaction.

Assumption II: *Once the reaction system has acquired enough energy to reach the top of the barrier, the probability of the trajectory leading to products is independent of all coordinates and momenta except for the reaction coordinates* q^* *and* p^*.

This assumption implies that memory of origin is lost after barrier passage. As a consequence, transition-state theory contains no information on the product. This is illustrated in the following example. The reaction of an oxygen atom and hydroxy radical can leave two different products

$$\begin{aligned} &O + OH \rightarrow H + O_2 && (a) \\ &O + OH \rightarrow HO_2 && (b) \end{aligned} \tag{4.137}$$

Both reactions proceed via the same reaction complex. Note that in (4.137*a*) the number of particles is conserved, whereas in (4.137*b*) the number of particles has decreased by 1 after reaction. In (4.137*a*), the reaction energy produced can be converted to relative translational energy, but in (4.137*b*) the HOO radical will become excited and decompose, unless a mechanism is provided that enables the vibrationally excited HOO radical to lose energy. Light emission or collisions with a third molecule may provide this possibility. If reactions such as (4.137) are studied, consecutive reaction steps have to be explicitly accounted for. The next chapter will look at the role of three body collisions.

Assumption III: *The energy is separable between the reaction coordinates* q^* *and* p^* *and the other variables. This implies that the total energy can be written as the sum of an energy term only dependent on* q^* *and* p^* *and a term that depends on other variables.*

These assumptions may be put into the following equations

$$v(q_1 \dots p_{3N-6}) = v(q^*, p^*) \qquad (a)$$

$$P_b(q_1 \dots p_{3N-6}) = P_b(q^*, p^*) \qquad (b)$$

$$\varepsilon = \varepsilon^* + \varepsilon' \qquad (c)$$

$$ds = ds' \, dq^* \, dp^* \qquad (d)$$

and

$$k_\sigma = \frac{1}{p.f._{\mathrm{AB}}^{\mathrm{rel}}}\left(\frac{1}{h^{3N-4}}\int_{6N-8}\cdots\int \exp^{-\varepsilon'/kT} ds'\right)\left(\frac{1}{h}\int\int dq^* \, dp^* \, v \times P_b e^{-\varepsilon^*/kT}\right) \quad (4.139)$$

The first term in the brackets is the partition function of the reaction complex excluding the reaction coordinate variables

$$pf^{\#} = \frac{1}{h^{3N-4}}\int ds' \, e^{-\varepsilon'/kT} \quad (4.140)$$

while the last term accounts for the reaction coordinate, and equals the frequency of barrier passage, v_b

$$v_b = \frac{1}{h}\int\int dp^* \, dq^* \, v \times P_b e^{-\varepsilon'/kT} \quad (4.141)$$

where P_b is the probability of barrier passage. The resulting expression for the rate constant, k_σ, becomes

$$k_\sigma = K^{\#} \, v_b \quad (4.142)$$

in which $K^{\#}$ is the equilibrium constant of the collision complex

$$K^{\#} = \frac{p.f.^{\#}}{p.f._{\mathrm{AB}}^{\mathrm{rel}}} \quad (4.143)$$

$p.f.^{\#}$ has to be computed for the geometry that corresponds to that of the complex, where coordinates p^* and q^* cross the barrier E_b. For this reason this particular complex is sometimes called transition-state or activated complex. Remember that the partition function $p.f.^{\#}$ depends on less phase space coordinates than $p.f.$, because the degrees of freedom of the reaction coordinate have to be excluded from $p.f.^{\#}$.

We will now evaluate v_b. Let δ be the distance that has to be crossed over the barrier in order that reaction occurs. Then

$$v = \frac{p^*}{\mu^*\delta} \quad (4.144)$$

and

$$v_b = \frac{1}{h}\int_0^\infty dp^* \int_{q^*}^{q^*+\delta} dq^* \left(\frac{p^*}{\mu^*\delta}\right) e^{-p^{*2}/2\mu kT} = \frac{kT}{h} \quad (4.145)$$

In order to evaluate ν_b, it is assumed $P_b = 1$ when the particle passes the barrier point q_b^*. A typical value for $kT/h \approx 10^{13}$ sec^{-1}.

With (4.145) the expression for k_σ becomes

$$k_{\mathrm{TST}} = \frac{kT}{h} K^{\#} \tag{4.146}$$

This is the transition-state expression for the rate of reaction. It is a rate, kT/h, multiplied by an equilibrium constant. Expression (4.146) has several interesting consequences.

It is interesting to note that (4.146) is the expected rate constant for the following sequence of elementary reactions

$$\mathrm{A} + \mathrm{B} \underset{k_1^f}{\overset{k_1^b}{\leftrightarrows}} \mathrm{AB}^{\#} \underset{k_2^f}{\rightarrow} \mathrm{P} \tag{4.147}$$

The reacting molecules A and B form activated complexes AB$^{\#}$ by collisions, and a fraction of these complexes crosses the activation barrier to form products P, with a rate constant k_2^f equal to $\nu_b = kT/h$.

The rate of product formation can be found by applying the steady-state approximation to AB$^{\#}$

$$\frac{d[\mathrm{AB}^{\#}]}{\mathrm{dt}} = k_1^f[\mathrm{A}][\mathrm{B}] - k_1^b[\mathrm{AB}^{\#}] - k_2^f[\mathrm{AB}^{\#}] = 0 \qquad (a)$$

or

$$[\mathrm{AB}^{\#}] = \frac{k_1^f}{k_1^b + k_2^f}[\mathrm{A}][\mathrm{B}] \qquad (b) \qquad (4.148)$$

and

$$\frac{d[\mathrm{P}]}{dt} = \frac{k_2^f k_1^f}{k_1^b + k_2^f}[\mathrm{A}][\mathrm{B}] \qquad (c)$$

Transition-state theory assumes that the number of collisions between A and B is much larger than the number of reactions. In reaction kinetics this means that the reaction step AB$^{\#}$ $\rightarrow$ P is rate limiting, while A+B and AB$^{\#}$ are considered to be equilibrated. Thus, with $k_2^f << k_1^b$

$$\begin{aligned} \frac{d[\mathrm{P}]}{\mathrm{dt}} &= k_2^f \frac{k_1^f}{k_1^b}[\mathrm{A}][\mathrm{B}] \\ &= k_2^f K^{\#}[\mathrm{A}][\mathrm{B}] \end{aligned} \tag{4.149}$$

Comparison with (4.146) shows that k_2^f equals $\nu_b = kT/h$.

According to (4.145) ν_b is non-activated, has non-Arrhenius behavior, and is only weakly temperature dependent. The reaction-rate activation energy depends on the enthalpy of the equilibrium constant $K^{\#}$. It can be evaluated from the partition functions of complexes and reagents as discussed in Section 2. Because this is a property of an equilibrium state it can also be evaluated directly using the analogue of relation (4.79*c*).

$$\begin{aligned} K^{\#} &= e^{-\Delta G_0^{\#}/RT} && (a) \\ &= e^{-(\Delta S_0^{\#}/R) - (\Delta H_0^{\#}/RT)} && (b) \end{aligned} \tag{4.150}$$

The gas constant R instead of Boltzmann constant k appears in (4.150). This implies that $\Delta G_0^{\#}$ is expressed in energy units per mole.

Substitution of (4.150b) into (4.146) gives

$$k_{\mathrm{TST}} = \frac{kT}{h} e^{(\Delta S_0^{\#}/R) - (\Delta H_0^{\#}/RT)} \tag{4.151}$$

Expression (4.151) has the form of the Arrhenius rate expression (4.130). The activation energy and the frequency ν_{eff} is found to be equal to (apply (4.131b))

$$\nu_{\mathrm{eff}} = e \frac{kT}{h} e^{\Delta S_0^{\#}/R} = \frac{kT}{h} e^{\Delta S_{\mathrm{act}}/R} \tag{4.152}$$

$$E_{\mathrm{act}} = \Delta H_0^{\#} + RT$$

The effective frequency is also called pre-exponential factor. The Arrhenius expression for the rate expressions can be written as

$$k_{\mathrm{Arr}} = \frac{kT}{h} e^{-\Delta S_0^{\#}/R} \cdot e^{-E_{\mathrm{act}}/RT} \tag{4.153}$$

The pre-exponential factor ν_{eff} can be considered as a frequency

$$\nu'_b = \frac{ekT}{h} \tag{4.154}$$

multiplied by

$$e^{\Delta S_0^{\#}/R} \tag{4.155}$$

$\Delta S_{\mathrm{act}} = \Delta S_0^{\#} + R$ is the activation entropy of the reaction. The larger the difference between the activation complex and that of the reactant, the larger the rate of the

TABLE 4.4. Data Needed for Calculating k_{TST} for the H + HBr Reaction

	H	H–Br	$(H–H–Br)^{\#}$
intermolecular distance (nm)	—	0.14	0.15 H–H 0.14 H–Br
vibrational frequencies (cm^{-1})	—	2650	2340 H–H stretch 460 bending mode 460 H–Br stretch [b]
mass (kg)	1.67×10^{-27}	1.338×10^{-25}	1.344×10^{-25}
moment of inertia (kgm^2)	—	3.3×10^{-47}	1.74×10^{-46}
partition functions:			
translation (m^{-3})	9.9×10^{29}	7.1×10^{32}	7.3×10^{32}
vibration	—	1	1.3
rotation	—	24.6	129.7
total	9.9×10^{29}	1.75×10^{34}	1.20×10^{35}

[a]Adapted from Laider (1987).
[b]This represents the reaction coordinate, accounted for by the factor kT/h.

reaction. When the activation energy is low, the rate of reaction is maximum when the entropy of the activated complex in the transition state is maximum. This is a typical "far from equilibrium" result. Generally, these reactions are favored that proceed with a minimum change in free energy $\Delta H_0^{\#} - T\Delta S_0^{\#}$.

Because $\Delta S_0^{\#}$ and $\Delta H_0^{\#}$ are thermodynamic properties, they can, in principle, be estimated from thermodynamics or computed by using the statistical-mechanical expressions discussed in section 2. The next section illustrates that it is essential to know the geometry of the transition state. However, often the transition state is not known and assumptions have to be made. Properties of the transition state can be deduced from transition-state rate theory when predictions are compared with experiments.

As an example, Table 4.4 contains the data necessary to calculate the rate constant k_T for the reaction

$$H + HBr \rightarrow (H–H–Br)^{\#} \rightarrow H_2 + Br \tag{4.156}$$

at room temperature. The transition-state expression for the reaction is

$$k_{TST} = \frac{kT}{h} \frac{(p.f.)^{\#}}{(p.f.)_H (p.f.)_{HBr}} e^{-E_0/RT} \tag{4.157}$$

The barrier energy corrected for the zero point energy contributions (4.163) is E_0 is 5 kJ/mol. Substitution of all numbers gives $k_T = 3.5 \times 10^9$ liter/mol.sec.

A comparison of the rate constants computed from transition-state theory (4.151) and that of hard-sphere collision theory (4.127), given in Table 4.5, clearly

TABLE 4.5. Comparison of Experimental and Theoretical Kinetic Parameters[a]

Reaction	Experimental E_{act} (kJ/mol)	Experimental $\log \nu_{eff}$[b]	Theoretical $\log \nu_{eff}$ transition-state theory	Theoretical $\log \nu_{eff}$ collision theory
$NO + O_3 \rightarrow NO_2 + O_2$	10.5	8.9	8.6	10.7
$NO + O_3 \rightarrow NO_3 + O$	29.3	9.8	8.6	10.8
$NO_2 + CO \rightarrow NO + CO_2$	132	10.1	9.8	10.6
$2NO_2 \rightarrow 2NO + O_2$	111	9.3	9.7	10.6
$2ClO \rightarrow Cl_2 + O_2$	0	7.8	7.0	10.4

[a]From Herschbach *et al.* (1956).
[b]Units of ν_{eff}: liter/mol. sec.

shows the effect of vibrational and rotational motion. It reduces the rate constant in these reactions by $10\text{–}10^3$. The explanation for this decrease is given in the next sections, where (4.151) will be applied to gas phase association and dissociation reactions as well as for some elementary surface reactions. Thanks to computational chemistry advances, transition states of reactions that are not too complex can be calculated. Using the calculated properties of transition states, the transition-state rate expression can be used as a predictive tool.

4.4. ASSOCIATION AND DISSOCIATION REACTIONS: ELEMENTARY REACTIONS ON SURFACES

Recombination of two atoms or molecule fragments is only possible if colliding molecules have a way to lose energy. The potential energy diagram for such a reaction is sketched in Figure 4.10. The relative energy has to be larger than E_b in order that a barrier for recombination can be crossed. If no kinetic energy of relative motion is lost, the colliding fragments will dissociate by return over the barrier and no molecule formation occurs. Only when energy is removed can the relative kinetic energy become low enough that the fragments bounce back at the inner part of the potential energy curve around r_b and association occurs.

In the gas phase, relative translational motion can be lost by inelastic collisions with a third particle in the short time interval that the distance between the atoms is smaller than r_b. When the colliding fragments are complex and have many internal degrees of freedom, relative translational energy can also be converted to internal vibrational or rotational energy. The internal excitations will lose their energy in consecutive collisions with other molecules. For adsorption on a surface, the lattice vibrations (or for metals, the metal electrons) may absorb energy and hence accommodate the energy that has to be released when a molecule adsorbs.

Similar considerations hold for dissociation and desorption reactions. The degree of freedom corresponding to the reaction coordinate of the dissociating molecule has to be activated. For a wide range of conditions it is found that the inter-conversion of translational energy into vibrational degrees of freedom and the reverse is very rapid. The rate constants can be evaluated using the transition-state expression for the rate constant as given in the previous subsection. The adsorption of transition-state theory rate constant for surface reactions will also be studied. Chapter 5 will return to deviations of the rate constant due to the absence of collisional deexcitation and excitation and study the conditions under which such non-equilibrium situations become important.

4.4.1. Dissociation and Desorption Reactions

> *The rate of dissociation is high for molecules that are weakly bound in the transition state. The Arrhenius activation energy is temperature dependent and* not *equal to the transition-state barrier energy.*

Consider the dissociation of a diatomic molecule AB into the atoms A and B

$$\mathrm{AB} \underset{k_1^f}{\overset{k_1^b}{\leftrightarrows}} \mathrm{AB}^{\#} \underset{k_2^f}{\rightarrow} \mathrm{A} + \mathrm{B} \tag{4.158}$$

According to transition-state theory, AB and AB$^{\#}$ are in equilibrium and the decomposition of AB$^{\#}$ into A and B is rate limiting. The rate constant of the overall process AB $\rightarrow$ A+B becomes

$$k_{\mathrm{diss}} = \frac{kT}{h} K^{\#} = \frac{kT}{h} \frac{p.f.(\mathrm{AB}^{\#})}{p.f.(\mathrm{AB})} \tag{4.159}$$

in which $p.f.(\mathrm{AB})$ stands for the partition function of the diatomic molecule

$$p.f. = p.f._{\mathrm{trans}}\, p.f._{\mathrm{vibr}}\, p.f._{\mathrm{rot}}\, e^{-U_0/kT} \tag{4.160}$$

(U_0 is the depth of the potential $V_{\mathrm{A-B}}$). For AB$^{\#}$ we have a similar expression, with E_b replacing U_0. However, we need to exclude from $p.f.(\mathrm{AB})^{\#}$ the degree of freedom representing the reaction coordinate, because the contribution of the latter is accounted for by the term kT/h. The reaction coordinate for the dissociation of AB is the distance r, determined by the vibration of the molecule. Thus, the expression for $p.f.(\mathrm{AB}^{\#})$ in (4.159) becomes

$$p.f.(\mathrm{AB}^{\#}) = p.f._{\mathrm{trans}}^{\#} \cdot p.f._{\mathrm{rot}}^{\#}\, e^{-E_b/kT} \tag{4.161}$$

The translational partition functions of AB and AB$^{\#}$ are the same, and the rotational partition functions cancel except for the r^2 terms, leaving the following expression for the rate constant:

$$k_{\text{diss}} = \frac{kT}{h} \frac{e^{h\nu/2kT}}{\left(1 - e^{-h\nu/kT}\right)^{-1}} \frac{r_b^2}{r_0^2} e^{-(E_b - U_0)/kT} \qquad (4.162)$$

When the molecule AB is strongly bonded, i.e., $h\nu$ is high, (4.162) reduces to

$$k_{\text{diss}} = \frac{kT}{h} \frac{r_b^2}{r_0^2} e^{-\Delta E/kT} \qquad (a)$$

$$\Delta E = E_b - U_0 - \frac{1}{2} h\nu \qquad (b) \qquad (4.163)$$

with: $h\nu >> kT$ (non-classical limit)

For weakly bonded complexes

$$k_{\text{diss}} = \nu \frac{r_b^2}{r_0^2} e^{-\Delta E/kT} \qquad (4.164)$$

with: $h\nu << kT$ (classical limit)

ΔE is again given by (4.163). Thus, (4.163) is an upper limit to (4.162). Indeed, it should be noted that in general the rate expression computed according to the transition state theory provides an upper limit to the exact reaction rate. The next chapter will return to this point.

Once we have an expression for the rate constant for the dissociation reaction, we also have the rate constant for the association reaction, k_{ass}, because the two are related through the equilibrium constant

$$k_{\text{ass}} = k_{\text{diss}} \cdot \overline{K}_{\text{eq}} \qquad (4.165)$$

where $\overline{K}_{\text{eq}}$ is given by expressions (4.87 and 4.88). It is interesting to compare k_{ass} from (4.165) with the expression derived from hard-sphere collision theory for k_σ in (4.127). The two are related in the following way

$$-\frac{dn_{\text{A}}}{dt} = k_\sigma \, n_{\text{A}} n_{\text{B}} \qquad (4.166)$$

n_{A} and n_{B} are concentrations

$$n_{\text{A}} = \frac{N_{\text{A}}}{V} \qquad (4.167)$$

The rate in change of the total number of molecules is

$$-\frac{dN_A}{dt} = \frac{1}{V} k_\sigma N_A N_B$$

$$= k_{ass} N_A N_B \tag{4.168}$$

Substitution of (4.163) and (4.87) into (4.165) gives (4.127).Transition-state and collision theories give the same result, because the association reaction of atoms A and B, which have no internal degrees of freedom has been considered and can be treated as hard spheres. Of course, the two theories would give different results for the association reaction between molecular fragments as for example in the reaction between CH_3 radicals to form ethane.

The Arrhenius activation energy is calculated from the expression

$$E_{act} = -k \frac{\partial}{\partial\left(\frac{1}{T}\right)} \ln k_{diss} \tag{4.169}$$

which gives (in units per mole)

$$E_{act} = \Delta H_{act} = \Delta E + RT \qquad (a)$$

$$\Delta S_{act} = R \ln e \frac{r_b^2}{r_0^2} \qquad (b) \tag{4.170}$$

and

$$\nu_{eff} = \frac{kT}{h} e^{\frac{\Delta S_{act}}{R}} \qquad (c)$$

Substitution of (4.170*b*) in (4.163) shows that the Arrhenius rate expression equals

$$k_{Arrhenius} = \frac{ekT}{h} \frac{r_b^2}{r_0^2} e^{-E_{act}/kT} \tag{4.171}$$

Expression (4.162) has been derived assuming r_b and r_0 to be independent of temperature. This is a reasonable approximation, but not completely correct.

For the non-classical limit, (4.163), $E_{act} = \Delta E + RT$ depends on the temperature, whereas in the classical limit, corresponding to a weak A–B bond, the activation energy equals ΔE. Comparison with Figure 4.11 shows that the activation energy indeed corresponds to the energy barrier the molecule AB has to surmount, however, starting from the zero-point vibrational energy, and not from the bottom of the potential well.

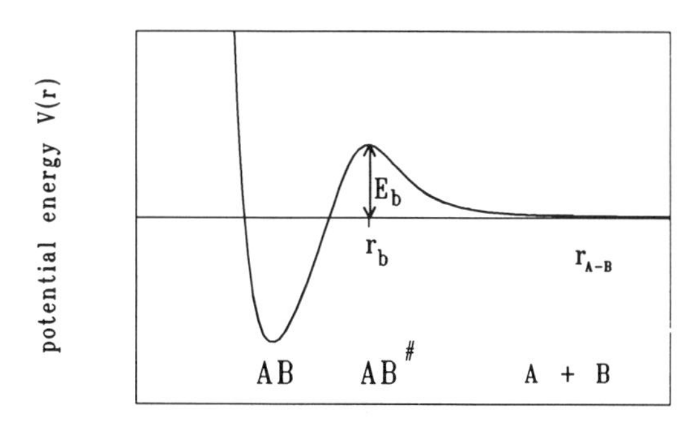

Figure 4.11. Potential energy diagram for the dissociation of a diatomic molecule AB.

The activation energy barrier for dissociation of a diatomic molecule is due to the centrifugal barrier. Remember that this contributes a term $p_a^2/2\mu r^2$ to the total energy expression. For an attractive potential of the form of $-C/r^6$, the van der Waals attraction potential, it equals (Rice, 1967)

$$r_c = \sqrt{1.728}\left(\frac{3c}{kT}\right)^{1/6} \tag{4.172}$$

For dissociation of a molecule larger than a diatomic molecule the transition-state theory expression of dissociation becomes

$$k_{\text{diss}} = \frac{kT}{h} \cdot \left(\frac{r_b}{r_0}\right)^2 \frac{(p.f.')^{\#}}{(p.f.')} e^{-(E_b - U_0 - \frac{1}{2}h\nu)/kT} \tag{4.173}$$

The partition functions now correspond to the internal degrees of freedom of the dissociating fragments as far as they do not cancel. The primes indicate that the partition functions have been taken with respect to E_b for $p.f.'^{\#}$ and to U_0 for $p.f.'$, implying that there is no exponential energy term in $p.f.^{\#}$, see (4.68).

For instance, the dissociation of ethane into two methyl radicals, $(p.f.')$ is due to the vibrational and rotational motion of the methyl fragments. When there is a large increase in rotational motion in the transition state

$$\left(\frac{r_b}{r_0}\right)^2 \frac{(p.f.')^{\#}}{(p.f.')} >> 1 \tag{4.174}$$

r_b is much larger than r_0 which assists the increase in the rotational motion. As can be seen from Table 4.1 the partition function of rotational motion is much larger

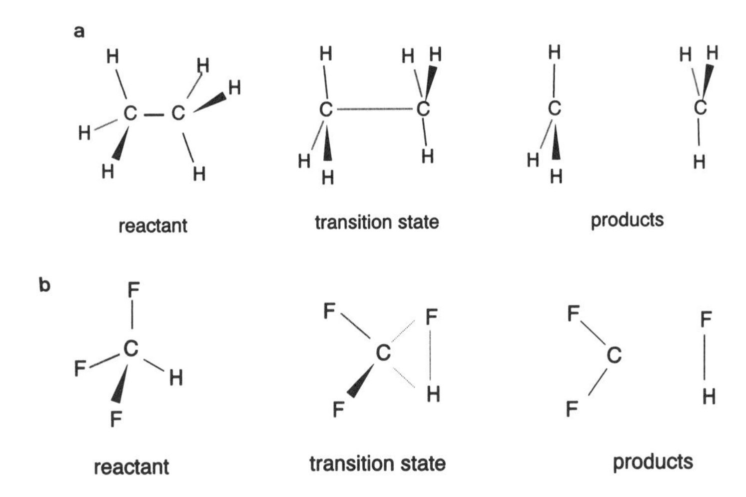

Figure 4.12. (a) Geometries of ethane in a loose transition state. (b) Geometries of fluoroform in a tight transition state.

than 1 (≈ 100). When motion is restricted (high frequencies), the partition function is close to 1. Condition (a) corresponds to a loose transition state. A tight transition-state complex corresponds to:

$$\frac{r_b^2}{r_0^2}\frac{(p.f.')^{\#}}{(p.f.')} \approx 1 \tag{4.175}$$

This will be illustrated with numerical examples for the dissociation of ethane and fluoroform. Figure 4.12a shows the computed geometries of ethane as reactant in its transition state and its dissociated products (Gilbert and Smith, 1990). In the transition state the carbon–carbon bond length is significantly increased and the CH_3 groups rotate almost freely. Typical pre-exponential factors are in the order of 10^{16}–10^{17} sec^{-1}.

Figure 4.12b shows the reactant, the transition state, and the products for the decomposition of fluoroform. They are the result of rigorous first principle calculations. The transition state is rather tight and the pre-exponential factor is in the order of 10^{13}–10^{14} sec^{-1}, very close to kT/h.

4.4.2. The Rate of Atomic Adsorption and Desorption

The free mobility of an adsorbed molecule increases during desorption. The increase is largest for atoms that desorb from an immobile state. In general surface diffusion will precede desorption. Two extra degrees of freedom during surface diffusion will enhance the entropy. Therefore the preexponent for desorption from the "mobile adsorption state" is considerably smaller than the one for desorption from the "immobile adsorption state."

This section and the following sections will explore the transition-state expressions for the rate constants of surface reactions. We start with the adsorption of atoms. The expression for the rate of adsorption according to hard sphere collision theory was covered. Expression (4.112) will be demonstrated as to how it can be rederived within the transition-state theory.

$k_{\text{ads}}^{\text{TST}}$ is defined by the rate expression

$$-\frac{dN_g}{dt} = k_{\text{ads}}^{\text{TST}}(1-\theta)\cdot N_g \tag{4.176}$$

where N_g is the number of gas-phase atoms in a volume V, and $(1-\theta)$ is the fraction of unoccupied sites on the surface. Commonly, concentrations are used rather than absolute numbers, and hence $n = N_g/V$ is defined as the concentration of gas-phase atoms. Anticipating that the transition-state expression for adsorbing atoms, which have no internal degrees of freedom, will be the same as that from hard sphere collision theory, the rate of change in the surface coverage is written as

$$\frac{d\theta}{dt} = \frac{V}{N_{\max}} k_{\text{ads}}^{\text{TST}}(1-\theta)\cdot n \tag{4.177a}$$

$$= k_{\text{ads}}^{\sigma}(1-\theta)\cdot n \tag{4.177b}$$

Substituting the transition-state expression (4.143) and (4.146) for $k_{\text{ads}}^{\text{TST}}$ gives for the rate constant of adsorption

$$k_{\text{ads}}^{\sigma} = \frac{V}{N_{\max}}\cdot\frac{kT}{h}\frac{(p.f.)^{\#}}{(p.f.)^{0}} \tag{4.178}$$

where $(p.f.)^0$ is the *p.f.* corresponding to the state before adsorption. This is the state of free three-dimensional translational motion

$$(p.f.)^0 = V\left(\frac{2\pi mkT}{h^2}\right)^{3/2} \tag{4.179}$$

The partition function in the transition state has to be computed for all degrees of freedom except the one corresponding to the reaction coordinate which is accounted for by the factor kT/h. The transition state, according to the hard sphere collision theory is free translational motion parallel to the surface and adsorption is assumed to be nonactivated, hence $(p.f.)^{\#}$ becomes

$$(p.f.)^{\#} = S \cdot \left(\frac{2\pi mkT}{h^2}\right) \tag{4.180}$$

Substitution of (4.179) and (4.180) into (4.178) gives

$$k^{\sigma}_{\text{ads}}\,(\text{hard sphere}) = \pi d^2 \cdot \frac{kT}{h}\,\frac{\dfrac{2\pi mkT}{h^2}}{\left(\dfrac{2\pi mkT}{h^2}\right)^{3/2}} \tag{4.181a}$$

$$= \frac{1}{4}\pi d^2 \left(\frac{8kT}{\pi m}\right)^{1/2} \tag{4.181b}$$

$$= \frac{1}{4}\bar{u}_s\,\pi d^2 \tag{4.181c}$$

This is indeed the expression (4.106) for the reaction-rate constant of hard spheres impinging on a surface, $\bar{u}_s$ is the average velocity of the atoms perpendicular to the surface and πd^2 is the effective surface area of an adsorbed atom ($S = N_{\max} \cdot \pi d^2$).

A molecule cannot be described as a hard sphere, and expression (4.181*c*) has to be replaced by

$$k^{\sigma}_{\text{ads}} = \frac{1}{4}\pi d^2\,\bar{u} \cdot \frac{(p.f.')^{\#}}{(p.f.')^{0}} \tag{4.182}$$

$p.f.'$ is the partition function of the internal degrees of motion. The sticking coefficient s_0 can now be computed using transition-state theory and is found to be

$$s_{\text{TST}} = \frac{k^{\sigma}_{\text{ads}}\,(\text{molecule})}{k^{\sigma}_{\text{ads}}\,(\text{hard sphere})} = \frac{(p.f.')^{\#}}{(p.f.')^{0}} \tag{4.183}$$

For a diatomic molecule, values of s_0 are typically between 0.001 and 1. When the molecule moves freely in the transition state and has unrestricted rotational motion $s_0 \approx 1$. However, when its rotational motion becomes restricted it may decrease by a factor 10^{-2}–10^{-3}. Theoretically it may decrease further if its translational motion becomes limited to a localized rigid transition state.

The desorption of an adsorbed atom from the surface will now be considered. One of the main differences between an atom adsorbed on a surface and one bonded

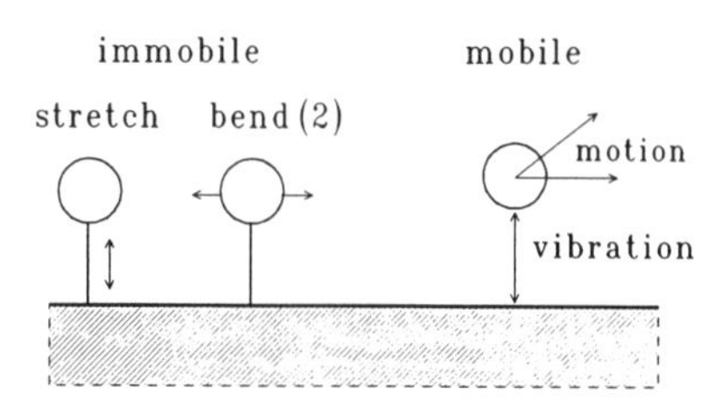

Figure 4.13. Degrees of freedom of an adsorbed atom.

in a molecule is the absence of rotational motion of the adsorption complex. The other important difference is that on a surface an atom may have a two dimensional diffusional motion in the state before desorption. Therefore, it is important to distinguish between desorption from a mobile ground state, in which the adsorbed atom moves freely over the surface, and desorption from an immobile ground state, in which the adsorbed atom can only vibrate around its equilibrium position. In both cases, the adsorbed atom has three degrees of freedom (Figure 4.13). When adsorbed in an immobile state it has one vibrational stretch mode with frequency μ_s perpendicular to the surface and two bending modes (ν_{bx}, ν_{by}) parallel to the surface. The transition state is then the state of free two-dimensional motion

$$k_{\text{des}}^{\text{imm}} = \frac{kT}{h} \frac{e^{\frac{1}{2}h\nu_s} e^{\frac{1}{2}h\nu_{b_x}} e^{\frac{1}{2}h\nu_{b_y}}}{\left(1 - e^{-h\nu_s/kT}\right)^{-1} \left(1 - e^{-h\nu_{b_x}/kT}\right)^{-1} \left(1 - e^{-h\nu_{b_y}/kT}\right)^{-1}} \times l^2 \left(\frac{2\pi mkT}{h^2}\right) e^{U_0/kT} \qquad (a)$$

$$= \nu_s \left(\frac{h\nu_b}{kT}\right)^2 \frac{2\pi mkT}{h^2} l^2 e^{\frac{U_0 + \frac{1}{2}h\nu_s + \frac{1}{2}h\nu_{b_x} + \frac{1}{2}h\nu_{b_y}}{kT}} \qquad (b) \qquad (4.184)$$

$$S = \frac{1}{4}\pi d_{\text{eff}}^2$$

$$U_0 < 0 \qquad (c)$$

Expression (4.184*b*) is the high temperature limit of (4.184*a*). This is valid when the surface vibrational frequencies are low, which is usually the case for atoms heavier than hydrogen. l^2 is the effective surface area that an adsorbed atom occupies. Its presence again derives from the two-dimensional translational partition function of the desorbed atom. A typical value of l^2 is 10^{-15} cm^2. Typical values for the surface translational partition functions are given in Table 4.6.

Generally the frequency ν_s may be expected to increase with more interaction energy of adsorbate and surface, U_0. One notes that a change of U_0 will be

TABLE 4.6 Two-dimensional Translation Partition Functions for Some Molecules at 500 K[a]

Molecule	$l^2 2\pi mkT/h^2$
H_2	33
CO	460
Cl_2	1200
Br_2	2600

[a]From Zhdanov *et al.* (1989).

counteracted by a change of ν_s. This is an example of the compensation effect. When an interaction parameter changes, the activation entropy and activation energy often change in an opposite way. This will be returned to in later examples.

The low temperature high frequency limit of expression (4.184*a*) becomes

$$k_{\text{des}}^{\text{imm}} = \frac{kT}{h}\left(\frac{2\pi mkT}{h^2}\right) l^2 \, e^{\frac{U_0 + \frac{1}{2}h\nu_s + \frac{1}{2}h\nu_{b_x} + \frac{1}{2}h\nu_{b_y}}{kT}} \tag{4.185}$$

Usually the bending modes have a lower frequency than the stretch mode, so an intermediate case may also exist

$$k_{\text{des}}^{\text{imm}} = \frac{kT}{h}\left(\frac{h\nu_b}{kT}\right)^2 l^2 \left(\frac{2\pi mkT}{h^2}\right) e^{\frac{U_0 + \frac{1}{2}h\nu_s + \frac{1}{2}h\nu_{b_x} + \frac{1}{2}h\nu_{b_y}}{kT}} \tag{4.186}$$

$$\text{intermediate case: } \frac{h\nu_b}{kT} << 1 \text{ and } \frac{h\nu_s}{kT} >> 1$$

The contribution due to bending frequencies counteracts the increase in rate due to the two-dimensional translational motion of the desorbing molecule. This will become clearer when the rate expression for mobile desorption is studied.

Measured values of the pre-exponent for desorption of metal atoms are typically between 10^{11}–10^{15} sec. Heavy metal atoms masses are large which implies that expression (4.184b) applies.

When the atom desorbs from a mobile state due to diffusional motion, the partition function of the bending vibration is replaced by one of diffusional motion

$$p.f._{\text{class}}^{\text{mob}} = \frac{1}{h^2}\int_{-\infty}^{+\infty}\int_{-\infty}^{+\infty}\int_{0}^{A}\int_{0}^{A} e^{-E/kT}\, dq_x\, dq_y\, dp_x\, dp_y$$

$$= \left(\frac{2\pi mkT}{h^2}\right) e^{-\left(\frac{U_0}{kT}\right)} \left(\int_0^A e^{\left(\frac{U_0}{2kT}\right)\cos\left(\frac{2\pi q}{d}\right)} dq\right)^2 \tag{4.187}$$

A is the effective surface distance in one dimension for an adsorbed molecule. A is small at high coverage, but may be very large at low surface coverage.

For the two dimensional potential this expression is used

$$V = \frac{1}{2} U_0 \left(2 - \cos \frac{2\pi x}{d} - \cos \frac{2\pi y}{d} \right) \tag{4.188}$$

V is the potential energy of the cubic lattice with periodicity $d/2\pi$. Substituting

$$\omega = \frac{U_0}{2kT} \tag{4.189a}$$

and $$\Theta = \frac{2\pi x}{d} \tag{4.189b}$$

gives $$p.f._{\text{class}}^{\text{mob}} = \frac{2\pi mkT}{h^2} A^2 e^{-2\omega} I_0^2(\omega) \tag{4.189c}$$

$I_0(\omega)$ is a modified Bessel function of the first kind.

$$\lim_{\omega \to \infty} I_0(\omega) = \frac{e^{\omega}}{(2\pi\omega^{1/2})} \tag{4.190}$$

The frequency of vibration is related to U_0

$$v_s = \left(\frac{U_0}{2md^2} \right)^{1/2} \tag{4.191}$$

In the low temperature limit ($\lim T \to 0$)

$$p.f._{\text{class}}^{\text{mob}} = \frac{A^2}{a^2} \left(\frac{kT}{hv_s} \right)^2 \tag{4.192}$$

In the classical limit the expression for desorption from the mobile state becomes

$$k_{\text{des}}^{\text{mob}} = v_s \left(\frac{d}{A} \right)^2 \left(\frac{hv_x}{kT} \right)^2 \cdot \frac{l^2 \cdot 2\pi mkT}{h^2} e^{U_0/kT} \tag{4.193}$$

The rate of desorption decreases proportionally to the effective surface area of the diffusing molecule

$$A^2 = \frac{S}{N}$$

$$= \frac{S}{N_{\max} \cdot \Theta} = \frac{\pi d_{\text{eff}}^2}{4\Theta} \qquad (a) \qquad (4.194)$$

and

$$d^2 = \frac{1}{4} \pi d_{\text{eff}}^2 \qquad (b)$$

where d^2 is the effective surface area of the desorbing molecule. Substitution of (4.194) results in the prediction that in the case of desorption from a mobile state, the desorption rate is a strong function of surface coverage

$$k_{\text{des}}^{\text{mob}} = \nu_s \Theta \left(\frac{h\nu_x}{kT} \right)^2 \cdot \frac{l^2 \cdot 2\pi mkT}{h^2} e^{U_0/kT} \qquad (4.195)$$

Finally, the quantum-mechanical partition function for mobile adsorption becomes

$$p.f._{\text{qm}}^{\text{mob}} = 2\pi \frac{A^2}{d^2} \omega e^{-2\omega} e^{-2K\omega} I_0^2(\omega) \frac{1}{(1 - e^{-K\omega})^2} \qquad (4.196)$$

with

$$K = \left(\frac{2h^2}{ma^2 V_0} \right)^{1/2} \qquad (4.197)$$

4.4.3. Rate of Molecular Desorption

The rate of molecular desorption may vary over a few orders of magnitude, depending on the rotation of the molecule in the transition state. When molecules go from a rigid adsorption state to a rotating transition state, the pre-exponential factor of the desorption rate constant increases by several orders of magnitude. The partition function of the internal vibration of a molecule remains close to one because the molecular frequencies are generally high compared to kT.

Chemisorption of CO on a transition metal can be considered the prototype of molecular chemisorption. A wealth of experimental information on CO is available. CO is chemisorbed with the carbon atom directed to the metal surface.

The center of mass of CO has a stretch mode ν_s^c perpendicular to the surface and two bending modes parallel to the metal surface (Figure 4.14). In addition the CO molecule has three internal degrees of freedom: A CO stretch frequency ν_s^m and

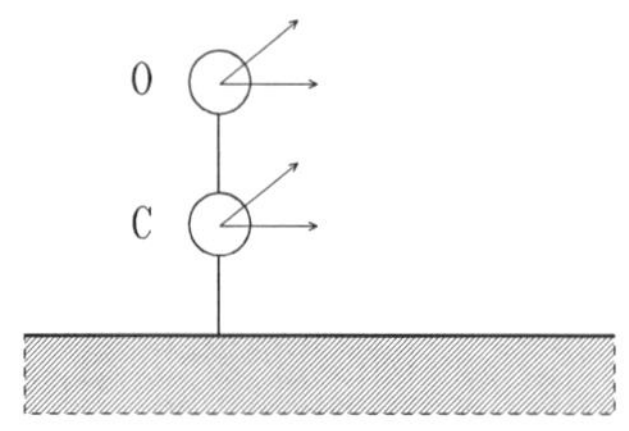

Figure 4.14. Vibrational modes of adsorbed CO along with a coordinate system to describe these modes.

two bending frequencies also parallel to the metal surface. They correspond to the two rotational degrees of freedom of the free molecule. For simplicity, the modes of the adsorbed molecule are described as if the molecule were in the gas phase. This is of course an approximation, but a useful one. The CO stretch-vibration frequency of linearly adsorbed CO (~ 2000–2070 cm^{-1}) is only slightly lower than that of gas phase CO (2143 cm^{-1}). In reality, the different modes are coupled in adsorbed CO, but it will be assumed that this coupling is zero in the following.

For the adsorbing molecule the center of mass motion of the CO molecule is chosen as the reaction coordinate. The resulting expression for desorption becomes very similar to expression (4.184) and the following, modified by the partition function quotient due to the change in internal motion

$$k_{\text{des}} = k_{cm}^{\text{des}} \frac{(p.f.')^{\#}}{(p.f.')_0} \tag{4.198}$$

with

$$(p.f.')_0 = p.f._{0\,\text{vibr}}^{\text{stretch}}(\text{AB})\; p.f._{\text{vibr}}^{\text{bending}'}(\text{AB})\; p.f._{\text{vibr}}^{\text{bending}''}(\text{AB}) \tag{4.199}$$

for the loose transition state

$$(p.f.')^{\#} = p.f._{\text{vibr}}^{\text{stretch}}(\text{AB})^{\#}\; p.f._{\text{rot}}(\text{AB})^{\#} \tag{4.200}$$

for the tight transition state

$$(p.f.')^{\#} = p.f._{\text{vibr}}^{\text{stretch}}(\text{AB})^{\#}\; p.f._{\text{vibr}}^{\text{bending}'}(\text{AB})^{\#}\; p.f._{\text{vibr}}^{\text{bending}''}(\text{AB})^{\#} \tag{4.201}$$

and k_{cm} given by an expression similar to (4.184).

When the molecule can be considered freely rotating in the transition state, this is called a loose activated complex. When rotation is restricted, it is considered a tight activated complex (Figure 4.15). Table 4.7 gives some values of rotational

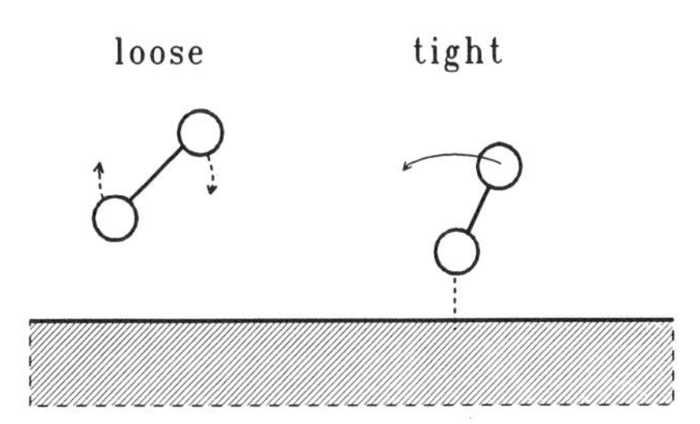

Figure 4.15. Loose and tight activated complex.

partition functions. The rate of desorption is maximum when the CO can be considered to rotate freely in the transition state

$$k_{\text{des}}^{\text{loose}} = k_{cm} \frac{8\pi^2 \mu r_0^2 kT}{h^2} \left(\frac{h\nu_{b'}^m}{kT} \right) \left(\frac{h\nu_{b''}^m}{kT} \right)$$

$$\frac{h\nu^c}{kT} < 1 \tag{4.202}$$

There is no contribution to the stretch frequency of CO because

$$\nu_s^m >> \frac{kT}{h} \tag{4.203}$$

μ is the relative mass of CO and r_0 the equilibrium distance. For CO at T = 500 K the rotational partition function equals 180. The rotational partition function enhances entropy of the transition state, which is favorable for desorption, but the effect is counteracted by the loss of the entropy contribution from the bending modes of adsorbed CO in the ground state.

From the results summarized in Table 4.8 it can be observed that the pre-exponent of desorption varies between 10^{13} and 10^{17}. On desorbing from an immobile

TABLE 4.7. Rotational Partition Functions for Some Molecules at 500 K.

Molecule	$8\pi^2 kTI/gh^2$
H_2	2.9
CO	180
Cl_2	710
Br_2	2100

TABLE 4.8. Activation Energy (kcal/mol) and Pre-exponential Factor (s^{-1}) for Desorption at Low Coverages[a]

	Low coverages	
System	E_a	ν
CO/Co(0001)	28	10^{15}
CO/Nu(111)	31	10^{15}
	37	10^{17}
	30	10^{15}
CO/Ni(100)	31	10^{14}
CO/Cu(100)	16	10^{14}
CO/Ru(001)	38	10^{16}
CO/Rh(111)	32	10^{14}
	32	10^{14}
CO/Pd(111)	34	10^{14}
	35	10^{15}
CO/Pd(100)	38	10^{16}
CO/Pd(211)	35	10^{14}
CO/Ir(110)	37	10^{13}
CO/Pt(111)	32	10^{14}
NO/Pt(111)	30	10^{15}
	27	10^{16}

[a]From Zhdanov *et al.* (1989).

state, expression (4.185) already gives a value of 10^{15} sec^{-1} for the pre-exponent. Free rotation of CO in the transition state (4.202) multiplies the pre-exponential value by another factor of 100.

4.4.4. Rate of Dissociation

The preexponent for the rate of dissociation of molecules adsorbed at a surface can be a few orders of magnitude less than that for desorption. Then the transition state for dissociation is a tight transition state.

Table 4.9 shows an important feature of dissociative adsorption. The pre-exponent of dissociative adsorption is generally orders of magnitude less than that for molecular adsorption. Assuming that desorption and dissociation occur from the same initial state

$$\frac{k_{\text{diss}}}{k_{\text{des}}} = \frac{(p.f.')^{\#}_{\text{diss}}}{(p.f.')^{\#}_{\text{des}}} e^{-(E_0^{\text{diss}} - E_b^{\text{des}})/kT} \tag{4.204}$$

Hence

TABLE 4.9. Pre-Exponential Factors for Dissociation of Adsorbed Molecules[a]

System	ν_{diss}/ν_{des}	ν_{diss}	E_{diss} (kcal/mol)
O_2 on Ag(110)	1.7×10^{-4}	1.7×10^{9}	7.8
O_2 on Ag(111)	1.7×10^{-6}	1.7×10^{7}	8.3
O_2 on Pt(111)	$9. \times 10^{-2}$	$9. \times 10^{11}$	2.5
N_2 on Fe(111)	$7. \times 10^{-6}$	$7. \times 10^{7}$	4.3
C_2H_6 on Pt(111)	$8. \times 10^{-4}$	$8. \times 10^{9}$	16.4
C_2H_2 on Pt(111)	$3. \times 10^{-1}$	$3. \times 10^{12}$	13.2
CH_4 on Rh film	$2. \times 10^{-3}$	$2. \times 10^{10}$	11.0

[a]From Campbell *et al.* (1991).

$$\frac{\nu_{\text{diss}}}{\nu_{\text{des}}} = \frac{(p.f.')^{\#}_{\text{diss}}}{(p.f.')_{\text{des}}} \tag{4.205}$$

When the $(p.f.')^{\#}$ of the dissociating molecule corresponds to that of the tight transition-state complex

$$(p.f.')^{\#}_{\text{diss}} \approx 1 \tag{4.206}$$

The frequencies of the complex are much higher than kT/h. Then

$$\frac{\nu_{\text{diss}}}{\nu_{\text{des}}} \approx 10^{-4} \tag{4.207}$$

when the transition state of the desorbing molecule is mobile and the molecule is freely rotating. This appears to be a reasonable description of the transition state for desorption (Table 4.8).

The conclusion that the transition state for dissociation of CO is tight is reasonable in view of the low activation energy of dissociation of adsorbed molecules. For instance, CO has a dissociation energy of approximately 215 kcal/mole in the gas phase. On a transition-metal surface CO dissociation typically has an activation energy of 20 kcal/mol. This enormous reduction in bond-breaking energy is mainly due to the stabilization of the fragment atoms by the metal-surface atoms when the C–O bond stretches. Loss in the CO interaction is compensated by the increasing interaction of the C and O atoms. This seems to be a very general picture for dissociation of adsorbed molecules. A tight interaction between the dissociating molecule and the metal surface corresponds to a low activation energy for dissociation. As will be discussed in Chapter 5, dissociative adsorption is rate limiting in many catalytic reactions. It is now understood why extremely low rates

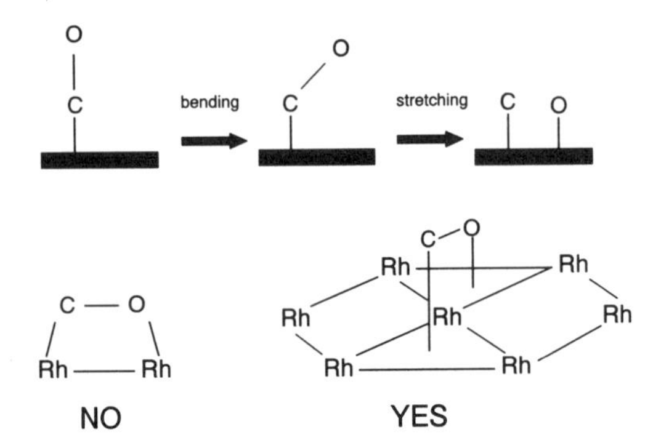

Figure 4.16. Transition state of dissociating CO in a CO molecule side bonded to a surface atom. (de Koster and van Santen, 1990).

of dissociative adsorption are often found. The tight transition-state complexes cause low pre-exponential factors and hence low reaction rates.

When a molecule dissociates, a different reaction coordinate has to be considered from the one used when the molecule desorbs. For the dissociating CO molecule the reaction coordinate is not the CO center of mass coordinate but the CO bond length. In the transition state the CO bond becomes stretched and CO is sideways bonded to a surface atom (Figure 4.16). After dissociation, the C and O atom are bonded in high coordination sites in sites opposite of the metal atom crossed by CO upon the dissociation (Figure 4.16).

Consider a surface where a molecule in a high coordination site has z surface-atom neighbors. Each can be used for dissociation when the CO molecule bends. The expression for the rate of dissociation becomes

$$k_{\text{diss}} = \frac{kT}{h}\, z\, \frac{(p.f.'_{\text{diss}})^{\#}}{(p.f.')_0}\, e^{-(E_b - U_0)/kT} \tag{4.208}$$

where z is the number of atoms surrounding the dissociating molecule.

The partition function of adsorbed CO in a mobile and immobile adsorption state has been discussed in the previous section. The partition function $p.f.'_{\text{diss}}(CO)^{\#}$, becomes

$$p.f.^{\#\prime}_{\text{diss}}(\text{CO}) = p.f.^{\#}_{\cdot\text{vibr}}(\nu_s^{cm})\, p.f.^{\#}_{\cdot\text{vibr}}(\nu_{b_x}^{cm})\, p.f.^{\#}_{\cdot\text{vibr}}(\nu_{b_y}^{cm})\, p.f.^{\#}_{\cdot\text{rot}_1} \qquad (a)$$

or (4.209)

$$= p.f.^{\#}_{\cdot\text{vibr}}(\nu_s^{cm})\, p.f.^{\#}_{\cdot\text{vibr}}(\nu_b^{cm})\, p.f.^{\#}_{\cdot\text{vibr}}(\nu_b^{cm})\, p.f.^{\#}_{\cdot\text{rock}}(\nu^{\text{CO}})\, p.f.^{\#}_{\cdot\text{rock}}(\nu_{r_y}^{\text{CO}}) \qquad (b)$$

Expression (4.209*a*) corresponds to the loose transition-state complex, and (4.209*b*) corresponds to the tight transition complex. In (4.209*a*) the molecule has been assumed to rotate in a plane parallel to the surface. In (4.209*b*), corresponding to a tight transition state, only rocking motions are possible.

Chapter 6 will cover how the transition states can be deduced from quantum-chemical calculations for surface reactions. Figure 4.17 shows the partition functions and corresponding pre-exponent of the dissociation of NO and CO as well as the recombination reactions of the separated atoms. The results confirm the tight binding nature of the transition states. Because of the wagging motion of the CO and NO bonds with respect to the surface normals of the molecularly adsorbed species, their partition functions are larger than that of the transition states or dissociated free atoms. The partition functions of the latter indicate very low entropies, due to the strong interaction with the surface. The pre-exponents for the recombination reactions differ nearly one order of magnitude from that of rate constants for dissociation.

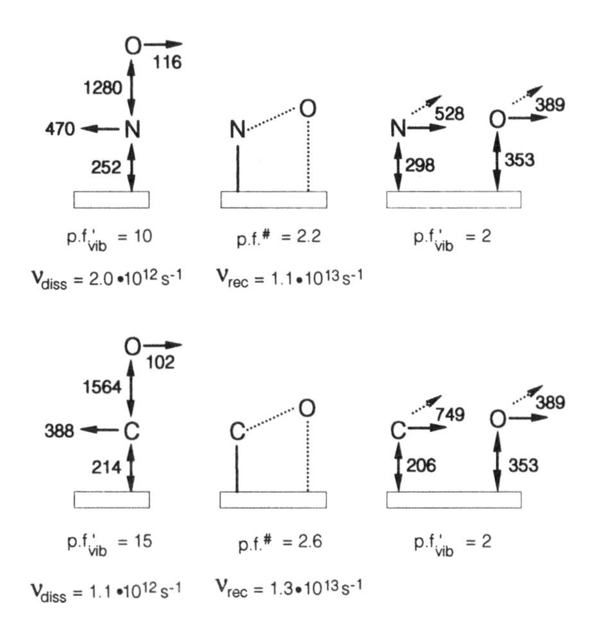

Figure 4.17. Groundstate and transition state partition functions for NO and CO on a Cu(111) surface computed according to density functional theory. The corresponding pre-exponents of the surface reaction rate constraints are also shown. (van Daelen, Newsam, and van Santen, 1994).

4.4.5. Surface Diffusion

The diffusion constant can be computed from transition-state theory. The pre-exponent depends on the square of the hopping length of a molecule.

As examined in the next chapter, transition-state theory is applicable when the rate-limiting step of a molecular process is controlled by a slow passage over a barrier. Not only reactions but also some diffusion processes are characterized by equilibration of a molecule on a surface site, in a cavity of a zeolite, and by movement of a molecule over a barrier (Figure 4.17). For diffusion, the relation between distance of motion and time is given by

$$r^2 = 4Dt \tag{4.210}$$

The surface sites, at a distance a, are separated by an energy barrier and diffusion is described by the "hopping" of adsorbates, grown one site to another, over the barriers. Therefore a is also called the hopping length.

$$a^2 = 4Dt_d \tag{4.211}$$

The diffusion constant then becomes

$$\begin{aligned} D &= \frac{1}{4t_d} a^2 \\ &= \frac{1}{4} k_d a^2 \end{aligned} \tag{4.212}$$

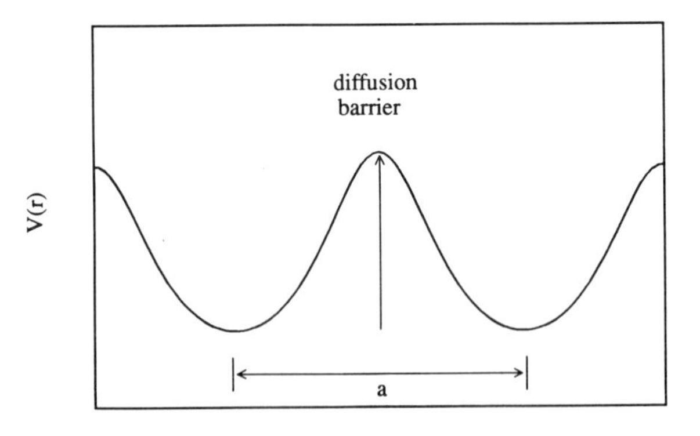

Figure 4.18. Barrier separating surface sites.

TABLE 4.10. Interval of Pre-exponential Factors for Some Elementary Surface Processes[a]

	Pre-exponential factor	
Process	Symbol and units	Value
Molecular desorption	ν, [s^{-1}]	10^{13}–10^{19}
Associative desorption	K_0, [cm^2s^{-1}]	10^{-4}–10^{4}
Monomolecular reaction	ν, [s^{-1}]	10^{12}–10^{13}
Surface diffusion	K_0, [cm^2s^{-1}]	10^{-2}–10^{-4}
Molecular adsorption	K_0, [cm^3s^{-1}]	10^{-10}–10^{-17}
Dissociative adsorption	K_0, [cm^3s^{-1}]	10^{-10}–10^{-17}

[a]From Zhdanov *et al.* (1989).

with

$$k_d = \frac{1}{2} z \frac{kT}{h} e^{-\varepsilon_a/kT} \frac{p.f.^{\#\prime}}{p.f.'} \tag{4.213}$$

This can be evaluated by well established means.

The diffusion constant for CO diffusion on a metal surface can be estimated with the understanding that during diffusion, the carbon bond remains bonded to the surface. Diffusion then results in loss of only one degree of freedom in the transition state, namely, its frustrated translational motion that has a frequency ν_T of ~30 cm^{-1}. Since this implies that $h\nu << kT$, k_d becomes

$$k_d = \frac{1}{2} z \, \nu_T \, e^{-\varepsilon_a/kT} \tag{4.214}$$

One important difference between diffusion and recombination processes is their activation energy. Their pre-exponents, however, appear to be related. The rate of associative desorption is proportional to the average surface area, whereas the diffusion constant relates to the square of the hopping length.

Table 4.10 gives a compilation of experimentally measured pre-exponential factors of surface processes. Note the variation in observed values.

5

MEDIUM EFFECTS ON REACTION RATES

5.1. INTRODUCTION

The medium in which a reaction takes place affects the reaction rate by the way energy is transferred to and from the reacting molecules, and by modifying their potential energy.

The previous chapter discussed the derivation of reaction-rate constants. This is for the case that the interaction of reactants with other molecules maintains temperature equilibrium between the degrees of freedom of reactants and its surrounding medium, considered to be a heatbath. As will be discussed in this chapter, this is a valid assumption only when the rate of the molecular events considered is slow compared to the rate of energy exchange between reactant molecules and the molecules in their environment.

Energy transfer and thermal accommodation form important considerations in the adsorption of molecules on a surface. Once adsorbed, the molecule may experience attraction or repulsion of other adsorbates on the surface, which may lead to the formation of "islands" of one type of adsorbate, or two-dimensionally ordered phases. Both phenomena have implications for the overall rate of surface-mediated reactions.

In addition to these statistical-mechanical effects, the medium often modifies the potential energy surface of reacting molecules. This is not only important in solvents, where it leads to solvation effects, but also in solids. When reactions occur in the narrow micropores of a zeolite, steric constraints of the zeolite channel will favor reaction paths that demand the least space.

5.1.1. The Chemical Potential of the Activated Complex

Solvation effects may change the interaction between reacting molecules. A polar medium will stabilize ionic or polar states. Reactions occurring in the micropores of solids may also experience steric constraints.

The presence of a medium affects the potential energies of the reacting molecules. For example, a polar transition state may become stabilized in a liquid or a solid with a high dielectric constant. Also reactions in which ions are formed will become energetically favored in polar media that stabilize the ions by hydration or coordination.

An example of a polar transition state on a solid is found for NO disproportionation. The zeolite NaY catalyzes the disproportionation reaction

$$4\,NO \rightarrow N_2O + N_2O_3$$

at a temperature below 0°C. The reaction proceeds according to

$$3NO \rightarrow N_2O + NO_2$$

$$NO + NO_2 \leftrightarrows [NO^+ - NO_2^-]^\# \rightarrow N_2O_3$$

Charge separation in vacuum would be endothermic by 5.45 eV. The polar reaction complex $[NO^+ - NO_2^-]$ is stabilized by the large electrostatic fields in zeolite cages due to the presence of highly charged cations. The zeolite plays a role analogous to that of a polar solvent. In this example, the medium facilitates the reaction by stabilizing a favorable transition state. The opposite situation may also occur as the following example illustrates.

Solvation can also change the potential energy of reacting molecules because of a stabilization of reactant or product fragments. An interesting example is provided in Figure 5.1 by the comparison of a computed reaction-energy profile of the nucleophilic substitution (a SN_2 reaction)

$$OH^- + CH_3Cl \leftrightarrows \left[HO\cdots\overset{H\;\;H}{\underset{H}{C}}\cdots Cl\right]^{\#}_{-} \rightarrow CH_3OH + Cl^-$$

In the gas phase, the reaction proceeds via three steps. OH^- can become weakly adsorbed on CH_3Cl (intermediate 1 in Figure 5.1) then proceed to the transition state 2. Cl^- can also become weakly bonded to CH_3OH (intermediate 3) before separation. In solution OH^-, as well as Cl^-, is solvated by water molecules. Upon reaction with the also solvated CH_3Cl molecule, water molecules around OH^- will

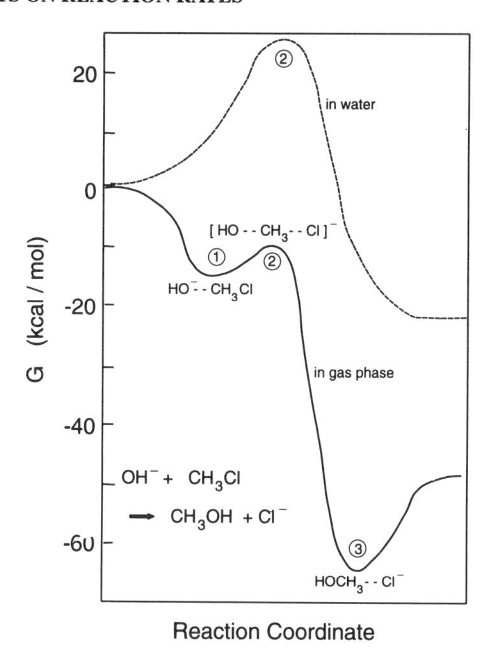

Figure 5.1. Potential energy curve along the reaction coordinate of the nucleophilic substitution reaction $OH^- + CH_3Cl \rightarrow CH_3 + Cl$ in the gas phase and in aqueous solution (after Evanseck *et al.*, 1985).

be replaced by CH_3Cl, a less polar molecule. In water the intermediates 1 and 3 will not be formed. In the gas phase OH^- or Cl^- will form a weakly bonded association complex, analogous to adsorbed states of interaction on surfaces. A significant contribution to the activation energy for a reaction in water is due to the rearrangement energy of the water molecules around the reactants. In the gas phase the association complex, which is only weakly bound, has to be decomposed.

In Figure 5.1 the potential energy curves have been computed by calculating the potential-energy surfaces of the reacting molecules for the isolated reactants and in the presence of a large number of water molecules. Note that the stabilization of OH^- or Cl^- by solvation removes the minima of the association complex and makes the reaction less likely than in the gas phase.

The difference in chemical potential between activation complex and reactant, $\Delta G^{\#}$ in a liquid may also depend on volume $\Delta V_0^{\#}$, as is evident from

$$\Delta G^{\#} = \Delta E_0^{\#} - T\Delta S_0^{\#} + P\Delta V_0^{\#} \tag{5.1}$$

The last term is especially important for loose transition-state complexes that occupy a large volume. The corresponding transition states will show a strong pressure dependence of the reaction rate in liquids.

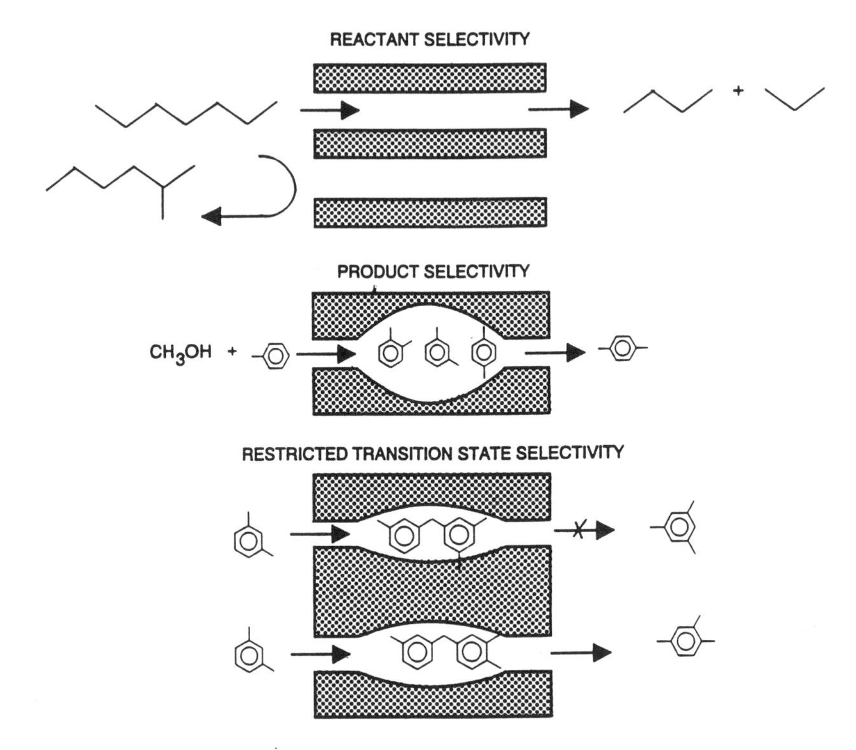

Figure 5.2. Application of zeolites results in selective conversion or formation of those molecules that fit in the zeolite channels.

Reactions that occur in the channels of the zeolites (Figure 5.2) can only proceed when the corresponding activated complexes do not generate a volume change $\Delta V_0^{\#}$ larger than allowed by the channel dimensions in which the reactions occur. Apart from this restricted transition-state selectivity, the microcavities of the zeolite may also influence the selectivity by excluding reactants that are too bulky. Although the larger molecules may be formed in cavities which offer sufficient space, diffusion of these molecules out of the zeolite will be slow or even impossible. The different kinds of selectivities in reactions in zeolites are illustrated in Figure 5.2.

5.2. ENERGY EXCHANGE AND TRANSITION-STATE THEORY

Associative reactions can only proceed if energy is removed from the product molecule. To this end, three-particle reactions are required. At low pressures, energy transfer to reactants can become rate limiting.

Under such conditions monomolecular reactions such as dissociation or isomerization behave as bimolecular reactions.

As previously mentioned, one of the conditions for the validity of transition-state theory is the rapid equilibration between the levels of the activated state and those of the ground state, implying that barrier passage is rate limiting. The different time scales involved depend on the rate of energy exchange with the surrounding medium. In a gas or a liquid, the rate at which molecules, and in particular reacting species in their transition state, gain or lose energy depends on the number of non-reactive collisions with molecules from the environment. On a surface, it depends on the amplitudes and frequencies of vibrating surface atoms.

Figure 5.3 illustrates how energy exchange influences the probability that two atoms form a diatomic molecule. If the atoms collide in the absence of any other atom or molecule, the chances are high that the atoms will bounce back and the molecule will not form. In other words, the collision complex formed at the moment of impact is so highly excited that it decomposes immediately. However, if the

a

	low pressure	high pressure
before collision		
moment of impact		
after collision		

b

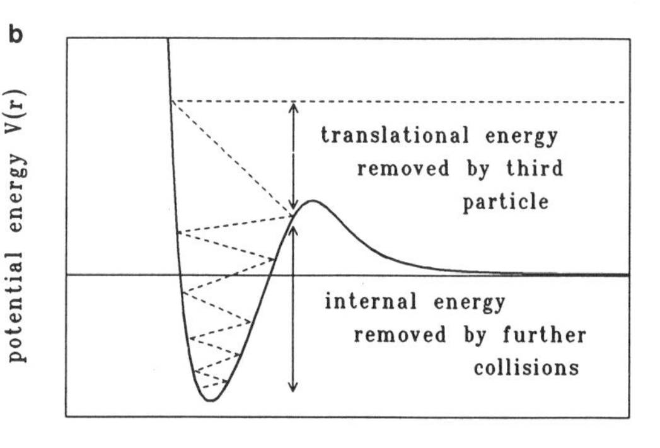

Figure 5.3. (a) The association reaction between two atoms can be successful only if the excess translational energy is removed by a third particle at the moment of impact. This leaves the newly formed molecule in a vibrationally and rotationally excited state. Deexcitation occurs through subsequent collisions with atoms or molecules from the surrounding medium. A potential energy diagram for this process is sketched in (b).

collision complex undergoes non-reactive collisions with one or more adjacent atoms or molecules, it can be sufficiently de-excited so rapidly that it becomes a stable diatomic molecule. The potential energy diagram in Figure 5.3 illustrates how the complex dissipates its excess energy. After removal of the translational energy by a medium molecule M, the complex A–A is left with an energy which falls inside the attractive well of the A–A interaction potential. It is still in a rotationally and vibrationally excited state. Further collisions with medium molecules de-excite the molecule to lower energy.

Here we will discuss the phenomena that result in chemical conversion reactions when transfer of energy and momentum become important. As demonstrated, the removal of translational energy (or momentum) is crucial in association reactions. The reverse process, collisional excitation of reactants is also important, for two types of unimolecular reactions, namely dissociation

$$A \rightleftarrows X + Y \tag{5.2a}$$

and isomerization

$$A \rightleftarrows A' \tag{5.2b}$$

Examples are the dissociation of molecular bromine, Br_2, the decomposition of sulfuryl, SO_2Cl_2, into SO_2 and Cl_2, in the isomerization of cyclopropane to propylene. All these reactions are elementary, and thus truly unimolecular.

In the early twenties, several scientists puzzled over how a reacting molecule in an unimolecular reaction could acquire sufficient internal energy to react. Jean Perrin proposed in 1919 that energy was provided by radiation from the walls of the reaction vessel. Frederick Lindemann, the later Lord Cherwell, strongly objected against this "Radiation Theory of Chemical Action" and presented an alternative view in 1921, which is still regarded as essentially correct. It explains a remarkable feature of unimolecular reactions: The rate of a unimolecular reaction is first order in the reactant at normal pressure, but may become second order at low pressure.

Figure 5.4 sketches the potential energy curves for dissociation and isomerization in one dimension. In both cases the molecule has to gain sufficient energy to undergo reaction. The Lindemann–Christiansen hypothesis, formulated independently by both scientists around 1921, says the molecule acquires the necessary energy by collisions with other molecules, after which it can either lose its energy in a subsequent collision, or cross the transition barrier to form products. The process is represented schematically by

$$A + A \rightleftarrows A^* + A$$

$$A^* \rightarrow \text{product} \tag{5.3}$$

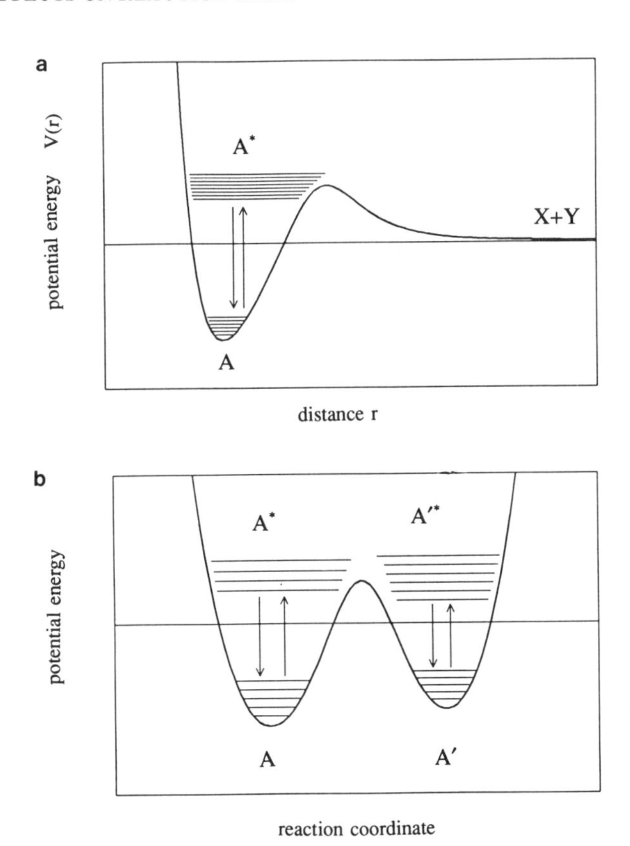

Figure 5.4. Potential energy diagrams for two unimolecular reactions: (a) dissociation and (b) isomerization.

In reality, as indicated in Figure 5.4, there are many excited states around the barrier that can all lead to reaction. According to activated complex theory, the rate is independent of the fate of the molecule once it has passed the barrier. The rate depends only on the concentration of the molecule in the activated state $[A_i^*]$ and the time necessary for passing the barrier, τ_i

$$r = \sum_i [A_i^*]k_i^f; \quad k_i^f = \frac{1}{\tau_i} \tag{5.4}$$

The index i labels the energy state of the molecule. When activated complex states A_i^* and ground state levels A_i are well equilibrated, (5.4) becomes equivalent to the one predicted by the transition-state theory.

The general form of the Lindemann mechanism is

$$\begin{gathered} A_j + M_k \xrightarrow{r^f_{1jkil}} A_i^* + M_l \\ A_i^* + M_l \xrightarrow{r^b_{1iljk}} A_j + M_k \\ A_i^* \xrightarrow{r^f_{2i}} \text{product} \end{gathered} \tag{5.5}$$

The indices label energy states of the molecules. The molecules M_k are either inert gas-phase molecules or the same molecules as A. Reactions (5.5) represent processes in which only energy and momentum is transferred between molecules. The rate equation for production of excited molecules is

$$\frac{d[A_i^*]}{dt} = \sum_{j,k} k^f_{1jki}[A_j][M_k] - \sum_l k^b_{1il}[[A_i^*][M_l] - k^f_{2i}[A_i^*] \tag{5.6}$$

The steady-state solution for the concentration of the activated intermediate is

$$[A_i^*] = \frac{\sum_{j,k} k^f_{1jki}[A_j][M_k]}{k^f_{2i} + \sum_l k^b_{1il}[M_l]} \tag{5.7}$$

Suppose the gas-phase molecules M undergo many mutual collisions such that their momenta and energies are equilibrated (the time scale of this process is the shortest). In this case

$$[A_i^*] = \frac{\sum_j k^f_{1ij}[A_j][M]}{k^f_{2i} + k^b_{1i}[M]} \tag{5.8}$$

As long as the distribution of ground state levels A_i remains close to equilibrium it is not necessary to distinguish between them and we may write

$$[A_i^*] = \frac{k^f_{1i}[A][M]}{k^f_{2i} + k^b_i[M]} \tag{5.9}$$

The rate of reaction becomes

$$\frac{d[P]}{dt} = \sum_i \frac{k_{1i}^f k_{2i}^f [M]}{k_{2i}^f + k_{1i}^b [M]} [A] \tag{5.10}$$

This is the general expression. Two limiting cases will now be considered.

In the limit of high pressure, relation (5.10) reduces to

$$\lim_{[M]\to\infty} \frac{d[P]}{dt} = \sum_i k_{2i}^f K_1^i [A] = k_T[A] \tag{5.11}$$

The rate constant becomes independent of pressure and can be computed using the transition-state theory expression. The reaction is first order in concentration $[A]$, as expected for a monomolecular reaction.

At lower pressures the reaction rate decreases and becomes dependent on $[M]$ with a low pressure limit

$$\lim_{[M]\to 0} \frac{d[P]}{dt} = \sum_i k_{1i}^f [M][A] \tag{5.12}$$

Thus, in the low pressure limit the rate is controlled by energy transfer between atoms A and M. The rate of reaction k_i^b of A_i^* states is so rapid that the rate of energy transfer to promote molecules from their ground state to the activated complex becomes rate limiting.

In case the gas consists only of molecules A, the low pressure limit expression becomes

$$\lim_{[A]\to 0} \frac{d[P]}{dt} = \sum_i k_{1i}^f [A]^2 \tag{5.13}$$

Note that the reaction is second order in A and that the rate has the form of a bimolecular reaction!

The general dependence of the rate constant k_R on M, as sketched in Figure 5.5 is

$$k_R = \sum_i \frac{k_{1i}^f k_{2i}^f [M]}{k_{2i}^f + k_{1i}^b [M]} \tag{5.14}$$

Thus, at low pressures the rate constant is low, because it is limited by energy exchange. At higher pressures, the rate constant approaches that predicted by the transition-state theory.

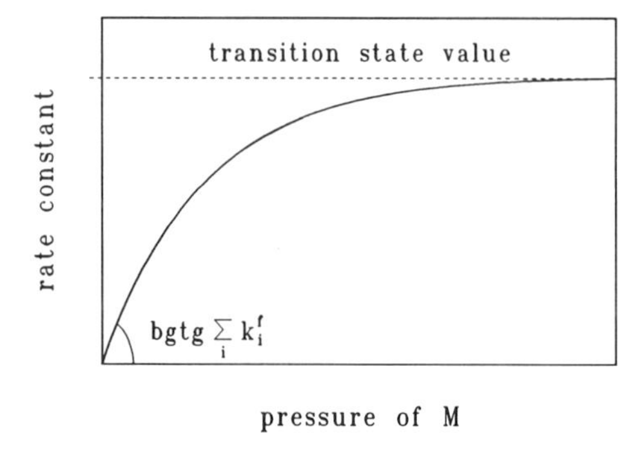

Figure 5.5. Rate constant of a dissociation reaction as a function of concentration of the inert gas phase molecules M.

The need to deactivate the molecules as soon as they are formed may lead to pressure dependent selectivities in reactions of the type

$$A + BC \begin{cases} \nearrow AB + C \quad \text{(path 1)} \\ \searrow ABC \quad \text{(path 2)} \end{cases} \tag{5.15}$$

Path 1 is favored at low pressure, whereas path 2 becomes more likely at higher pressure. In this scheme, path 1 denotes a reaction path where the total number of free particles remains the same so that energy can be converted into relative translational energy. Path 2 is an association reaction that can only occur when collisions with a third particle remove energy.

In the following gas phase radical reactions

$$O + OH \begin{cases} \nearrow O_2 + H \\ \searrow OOH \end{cases} \tag{5.16}$$

O_2 formation occurs at all pressures, but OOH forms only at a higher pressure.

A molecule adsorbing on a metal surface also has to lose its energy. In vacuum this can only occur by a loss of translational energy to the lattice vibrations of the solid. This process is called thermal accommodation.

Later on, this chapter will give the rules that control the distribution of energy between translational and vibrational modes. It will appear that at atmospheric pressures it is often correct to consider thermal equilibration as a rapid process so that the rate expressions can be computed in the limit of high pressure (5.11).

To summarize, at low pressures the rate of energy transfer becomes rate limiting, and the reaction rate constant is lower than the one predicted by transition state theory. However, at high pressures, when the number of energy and momentum exchanging collisions becomes very large, the reaction-rate constant again decreases. This phenomenon is explained by Kramers' reaction-rate theory, discussed in the next section. Thus, in general the transition-state theory gives an upper limit for the reaction rate.

5.3. KRAMERS' REACTION-RATE THEORY

The rate equation predicted by the transition-state theory is an upper limit. If energy transfer is slow, the rate decreases. In dense media, the rate also decreases when too many collisions between the reacting molecule and the medium occur, because the molecule may now bounce back from the transition barrier into its initial state before reaction. A high viscosity diminishes the reaction rate.

In 1940, Hendrik Kramers rigorously derived the conditions under which transition-state theory is valid. Rate processes are determined by the rate of passing the transition barrier and excitation or deexcitation processes of a reacting system to those energetically activated states from which barrier passage becomes possible. Because the overall rate is small and the number of collisions with medium molecules is large, the general statistical dynamical theory that treats this considers the reaction of the molecule to occur in a stochastic, randomly fluctuating force field. The concept of Brownian motion in liquids is an example of such a theory: A macroscopic particle moves as the result of many collisions with other molecules in the liquid, resulting in a diffusive motion of the particle according to the law

$$\langle r^2 \rangle = 6Dt \tag{5.17}$$

A similar diffusion rate is also valid for an atom moving over a surface (with the six in (5.17) replaced by four), when the number of energy-exchanging events of the moving atom with the atoms of the metal surface is large compared to the overall diffusion rate.

When a particle moves through a liquid, it experiences a frictional force. This is the origin of viscosity. This property is caused by random collisions between the particle and medium molecules. Stochastic theory again applies if the overall motion is slow compared to the number of collisions.

The Stokes–Einstein equation gives the relation between viscosity and diffusion constants

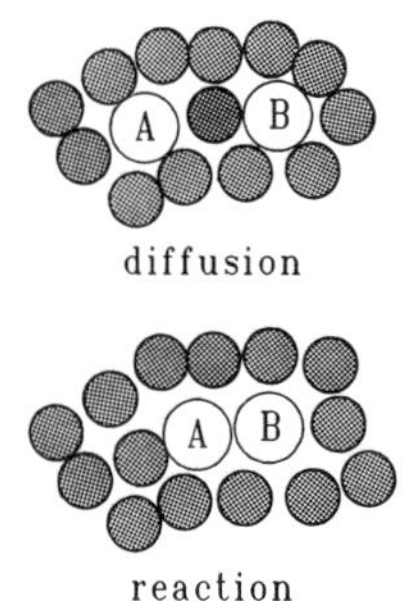

Figure 5.6. Chemical reactions in dense media require that the reactants come together through diffusion.

$$D = \frac{kT}{6\pi\eta R} \tag{5.18}$$

in which k is Boltzmann's constant, T is the temperature, η the viscosity and R is the radius of the particle that moves through the liquid.

A reaction between spherical atoms or molecules A and B in a solution occurs in two steps (Figure 5.6) namely diffusion of A and B toward each other, followed by the actual reaction

$$A + B \overset{k_d}{\rightleftarrows} AB \overset{k_{\text{chem}}}{\rightleftarrows} \text{products} \tag{5.19}$$

where k_{chem} is the rate constant in the absence of any medium effect, and the rate constant of diffusion is given by

$$\begin{aligned} k_d &= 4\pi(D_{\text{A}} + D_{\text{B}})\,(R_{\text{A}} + R_{\text{B}}) \\ &= \frac{2kT}{3\eta}\,\frac{(R_{\text{A}} + R_{\text{B}})^2}{R_{\text{A}}R_{\text{B}}} \end{aligned} \tag{5.20}$$

In case A and B are the same, (5.20) becomes

$$k_d = \frac{8kT}{3\eta} \tag{5.21}$$

and the rate constant for diffusion depends only on the viscosity of the medium and on the temperature. For water at room temperature, (5.21) yields a value of $k_d = 7 \times 10^9$ 1/mol.sec.

The overall rate constant for (5.19), k_r, can be written as

$$k_r = \frac{k_{\text{chem}}}{1 + \dfrac{k_{\text{chem}}}{k_d}} \tag{5.22}$$

This equation has two limiting cases. When k_{chem} is much larger than k_d, the overall rate constant equals k_d, and the reaction is fully controlled by diffusion. This may happen in dense liquids or gases at high pressure. The other extreme occurs if $k_{\text{chem}} << k_d$, in this case there is no effect of the medium on the reaction. In intermediate cases, the reaction is said to be in partial microscopic diffusion control.

The separation of a process as (5.19) in diffusion and unperturbed chemical reaction is not entirely correct, as the energy exchange between the reacting complex and the medium may de–excite the transition state and lower the probability that the system crosses the energy barrier to form products. This possibility is discussed in the following.

Figure 5.7 sketches the potential for one reaction coordinate of a reacting system. Such a diagram is typical for an isomerization reaction. The frequencies ω_a, ω_b, and ω_c correspond to the frequency of a particle moving in the corresponding potentials. In order to move from position a to b, stochastic collisions have to occur such that energy is transferred and the system moves to the maximum of the energy barrier. Energy transfer from the ground-state energy levels to those of the excited state of the activated complex can be considered as the ladder step process of Figure 5.7. Movement over this ladder under the influence of environmental collisions can be viewed as a diffusive motion through energy space, characterized by a frictional or viscosity constant η. If the friction is small, then

$$k_r \propto \frac{\eta}{kT} k_T < k_T \quad \left(\frac{\eta}{kT} < 1\right) \tag{5.23}$$

with k_T the rate constant according to the transition-state theory (left part of curve in Figure 5.8). Expression (5.23) is valid in the case that friction is small.

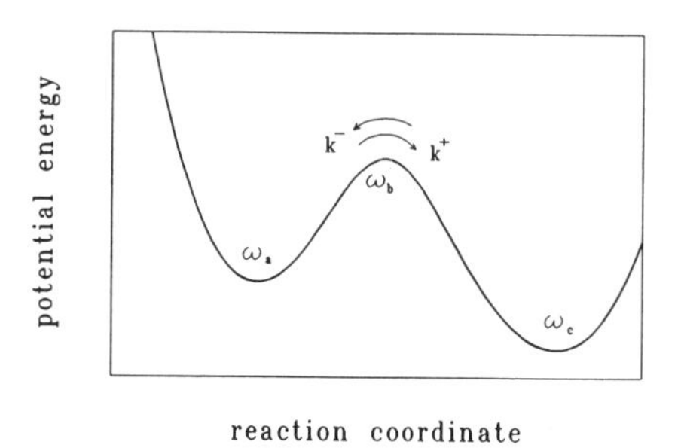

Figure 5.7. Reaction-rate diagram for a reaction in a dense medium.

The proportionality of the rate to the friction constant is the analogue of (5.18), the low pressure limit for the unimolecular rate constant. In this limit the rate limiting step is collisional energy transfer. The overall reaction rate is small compared to the one predicted by the transition-state rate expression.

If friction is important

$$\eta >> \omega_b \rightarrow k_r = \frac{\omega_b \omega_a}{2\pi\eta} e^{-E_b/kT} \tag{5.24}$$

This rate expression, responsible for the right side of Figure 5.8, is important in a liquid with a high viscosity. Use of the Stokes–Einstein relation (5.18) converts (5.24) to

$$k_r = \frac{R^{\#}D}{kT} \omega_b \omega_a e^{-E_b/kT} \tag{5.25}$$

Note that k_r in (5.25) depends similarly on R and D as the diffusive recombination rate constant in (5.20). During passage over the barrier the particle responds to a friction with the medium, due to the viscosity of the latter, that slows down the rate. D is the diffusion constant for the movement over the barrier.

The general equation for this process is given by the Langevin equation, which, in one dimension is

$$\ddot{x} = -\frac{1}{m}\frac{\partial U}{\partial X} - \eta\,\dot{x} + \frac{1}{m}\xi(t) \tag{5.26}$$

x is the reaction coordinate, m the reduced mass, and U the potential. Expression (5.26) is actually Newton's force equation to which two terms have been added: A

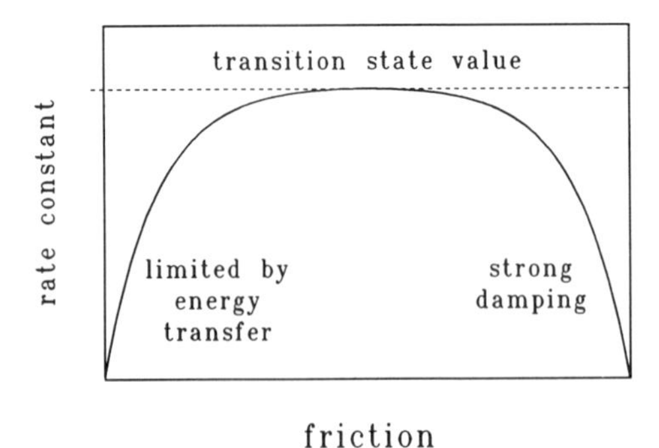

Figure 5.8. Relation between the rate constant and the frictional constant.

frictional term that depends on the rate of the reaction coordinate causing the rate to slow down due to energy dispersion towards the medium, and a fluctuating force $\xi(t)$. The fluctuation force depends on the temperature and results from the stochastic motion of the colliding molecules. It restores the equilibrium in the system. Expression (5.26) provides a link between chemical kinetics and molecular dynamics.

The time average of the fluctuating force has to compensate the loss in energy due to friction

$$< \xi(t)\ \xi(t') > = 2kT\ \eta\ m\ \delta(t - t') \tag{5.27}$$

where δ is the delta function, defined such that

$$\lim_{\in \to 0} \int_{t-\in}^{t+\in} ds\ \delta(t - t') = 1$$

$$\lim_{\in \to 0} \int_{t+\in}^{\infty} ds\ \delta(t - t') = 0$$

$$\lim_{\in \to 0} \int_{-\infty}^{t-\in} ds\ \delta(t - t') = 0 \tag{5.28}$$

Microscopic methods enable an estimate of η. Kramers' results follow in the limit of low and high friction.

The dynamic motion due to rapid energy exchange for the desorption of Xe atoms from a Pd(100) surface will be illustrated. Figure 5.9a shows the rate of Xe desorption as predicted according to transition-state theory. Figure 5.9b compares computed molecular-dynamics rates and the transition-state rates. The open data points are the computed desorption rates for Xe atoms that are allowed to readsorb once they have passed the transition-state barrier. The filled data points ignore the possibility of readsorption. The open data points, computed from the more exact theory, always remain lower than the transition-state result. Transition-state theory and molecular dynamics predict very similar rate constants for the desorption of xenon from palladium.

It is interesting to follow the microscopic motion of the Xe atom at temperatures lower (Figure 5.10a) and higher than the temperature of desorption (Figure 5.10b). The motion of the Pd atoms kicks the Xe atoms into different directions, parallel or perpendicular to the surface. Only at higher temperature is energy transfer between metal atoms and Xe atoms such that the motion of the Xe atom

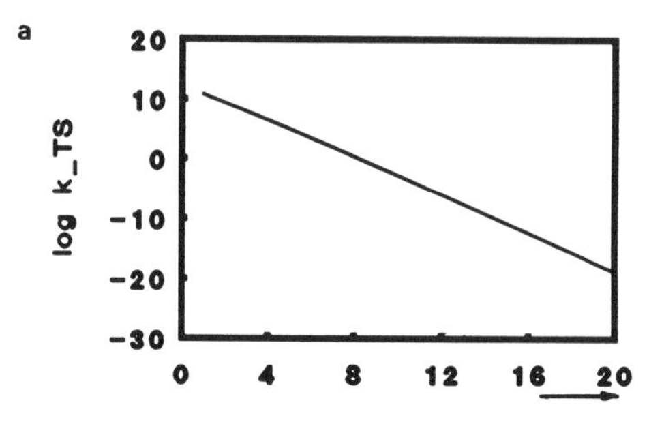

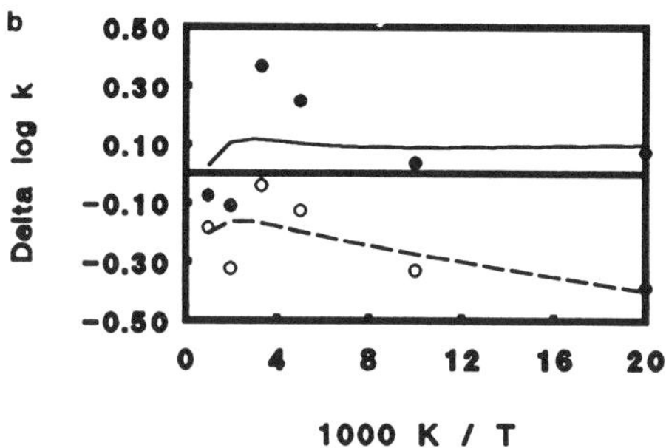

Figure 5.9. (a) Arrhenius plot of the rate of xenon desorption from a palladium (100) surface as calculated with transition-state theory. (b) Difference between the rates from a molecular dynamics simulation based on expression (5.26) and those calculated according to transition-state theory. The open data points include the possibility of readsorption which is ignored in the simulations corresponding to the filled points (from A.P. J. Jansen, 1992).

can overcome the attractive energy well of the Xe atom and the metal surface. Below the desorption temperature the only motion possible for the Xe atoms is random diffusion over the metal surface.

Using the solutions of the Langevin equation for diffusional motion of CO on a Ni(111) surface, Dobbs and Doren have determined the hopping length, a of (4.211) that is needed to compute the surface-diffusion constant from the transition-state expression k_d given by (4.213). They found that for CO dissociation on Ni the usual assumption of random, near neighbor hops are not correct. By following the individual trajectories of the CO molecules hopping over the surface, they computed the fraction of molecules that crossed the surfaces that defined the boundaries of surface atoms. As illustrated in Figure 5.11, at 200K only 30% of the trajectories consisted of nearest-neighbor surface atom crossings. Because of the average larger hopping length, the computed surface diffusion constant is a factor 10 higher than the value estimated from (4.211).

The transition-state theory assumption implies that the diffusing molecule has no excess energy once it has crossed the transition-state barrier. It loses its energy on a time scale short compared to the diffusion time τ_d. The molecular dynamics calculations show that for the CO/Ni system, energy exchange between CO and the metal surface is slow, giving an increase of the diffusion rate compared to the transition-state result. According to the calculated results, diffusion of CO on Ni

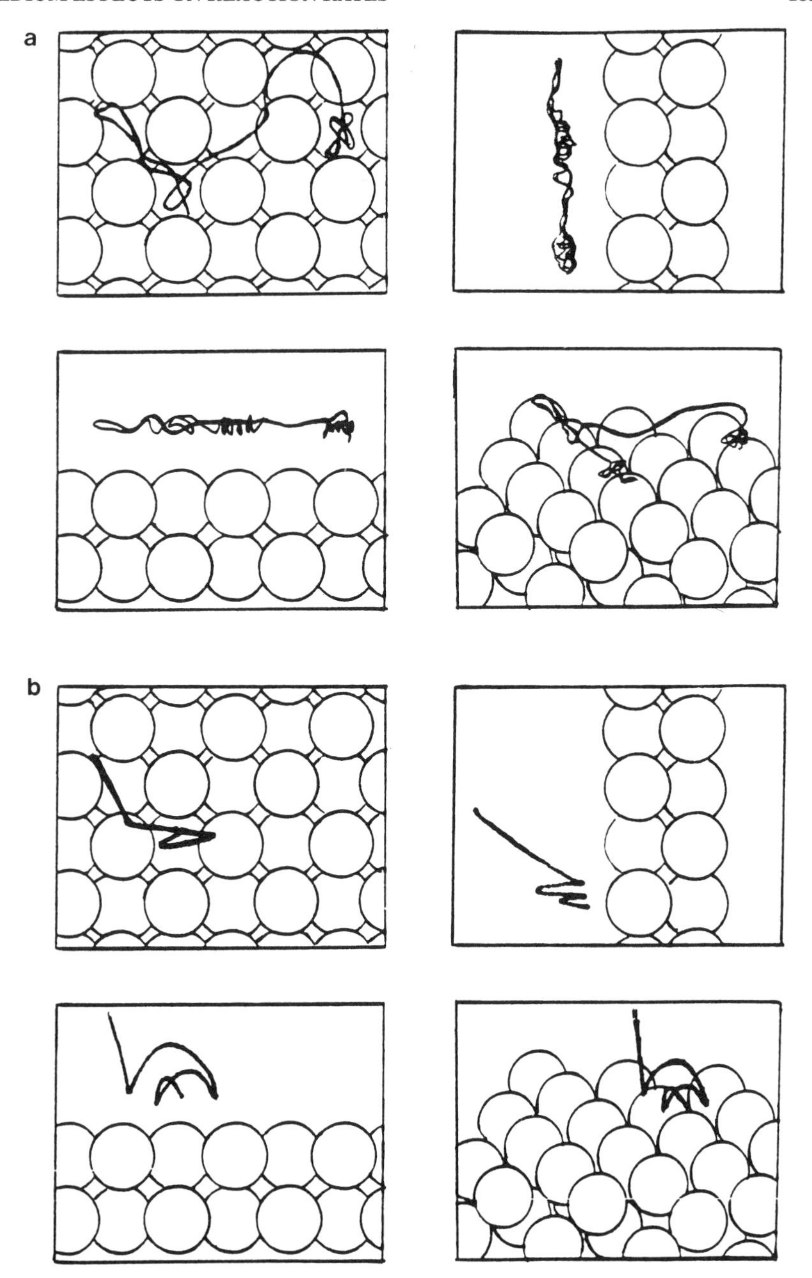

Figure 5.10. Molecular dynamics simulation of the diffusion of a xenon atom on a Pd(100) surface, (a) below and (b) above the desorption temperature. Shown are a top view, two side views, and a three dimensional perspective (courtesy of A.P. J. Jansen, 1992).

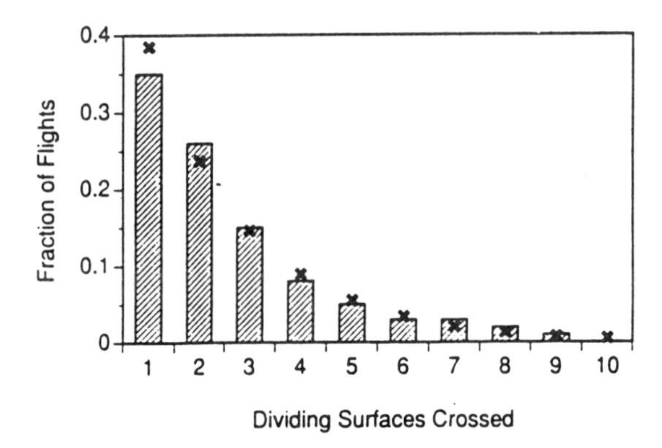

Figure 5.11. Distribution of dividing surfaces crossed per flight for trajectories at 200K. Only one-third of flights end after a single dividing surface crossing. (Dobbs and Doren, 1992).

occurs near the limit of low friction. Once the molecule has gained energy, this energy is only slowly dissipated to the metal surface. As a consequence it obtains a rather long hopping distance.

In ultra-high vacuum experiments low friction was indeed observed for the rate of energy dissipation of diffusing atoms. The oxygen atoms generated by dissociation of adsorbed molecular oxygen on an Al(111) surface move about 8 nm in opposite directions before they become adsorbed (Brune *et al.*, 1992). In surface dissociation the activation-energy barrier is high compared to the energy of chemisorbed oxygen atoms. The excess energy generated upon passage of the barrier has to be released. In the absence of gas phase collisions, the slow dissipation of energy to the surface implies that hot atoms form which diffuse over long distances before they become equilibrated with the surface.

Kramers has rigorously derived the conditions under which transition-state theory applies. One is that the different time regimes already discussed have to be separable. Another condition is that the thermal excitation energy, kT, is considerably smaller than the energy barrier between reactant and product

$$E_b > 5kT \tag{5.29}$$

As long as this condition holds, the states of the reacting complex close to the transition-state barrier can be considered equilibrated. Only a few particles escape over the barrier, and hence equilibrium is maintained.

The third condition has to do with friction. When friction becomes too large, diffusion over the barrier will be hindered. On the other hand, when friction is too small, energy transfer becomes rate limiting. The condition for the validity of the transition-state rate expression is

$$\frac{kT}{E_b} < \frac{\eta}{\omega_b} < 1 \tag{5.30}$$

The general relation between η and k_r is sketched in Figure 5.8. Deviations from the transition-state expression always have to be considered carefully when activation energies are very low. The use of the transition-state expression in xenon desorption from the metal surface is appropriate because of the relative high value of the desorption energy compared to kT. Transition-state theory is not valid for surface diffusion of CO because of the low activation energy. Of course, quantitative methods are needed to predict a priori the transition state. This will be demonstrated in Chapter 6.

5.4. RATE OF VIBRATIONAL ENERGY TRANSFER BETWEEN GAS MOLECULES

The transfer of energy in collisions between molecules with several degrees of freedom is highly efficient, reactions of such molecules have a rate almost equal to that predicted by transition-state theory. The efficiency for energy transfer between atoms and molecules from translational to vibrational energy decreases with increasing collision time and is most efficient for molecules with a small mass.

The previous sections demonstrated that reaction rates may be limited by intermolecular energy transfer at low pressure. As a consequence, transition-state theory overestimates the rate constant, and the appropriate expression for the rate constant should contain energy-transfer parameters and non-equilibrium factors.

Experiments where the reactants are present at low concentrations in a non-reactive medium, often called bath gas, have contributed significantly to our understanding of intermolecular energy transfer. Collisions with the bath gas increase the rate constant, sometimes even to the value it has for the reaction at high pressure. However, not all bath gases are equally efficient in transferring energy. A parameter called collision efficiency, β_c, accounts for this:

$$k_r = \beta_c k_T \quad (\beta_c \leq 1) \tag{5.31}$$

where k_r is the rate constant for the reaction of reactants in a non-reactive bath gas and k_T the one predicted by transition-state theory. The collision efficiency β_c is related to the average amount of energy transferred per collision, ΔE. In the strong collision limit where $\beta_c \approx 1$ (or even higher), the transferred energy is at least equal to the thermal energy, $\Delta E \approx kT$. The collision efficiency decreases rapidly with increasing temperature. In the limit of weak collisions the following empirical relation between β_c and ΔE may be used

TABLE 5.1. Energy Transfer in Various Unimolecular Reaction Systems in a Surrounding Medium (the Bath Gas) [a]

ΔE (kJ/mol) for energy transfer in various unimolecular reaction systems with bath gases M

M	$I + NO_2 \rightarrow INO_2$ 300 K	$CH_3NC \rightarrow CH_3CN$ 554 K	$C_2H_5NC \rightarrow C_2H_5CN$ 504 K	$C_7H_8 \rightarrow C_6H_5CH_3$ 300 K	$CH_3 + CF_3 \rightarrow C_2F_3H_3$ 300 K
He	1.2	2.1	2.5	1.7	2.6
Ne	3.3	2.8	2.9	3.5	2.6
Ar	3.3	2.8	5.0	3.8	1.9
Kr	3.3	2.1	6.3	5.7	1.9
Xe	4.4	1.8	5.7	6.1	2.6
H_2	0.8	2.1	2.4	2.9	4.5
D_2	1.0	2.3		3.4	2.6
N_2	4.1	4.6	6.1	3.1	4.5
CO	4.4	5.5		5.2	
O_2	6			4.6	
N_2O	8			7.3	
CO_2	8	9	16	9.8	6.4
CH_4	3.7	11	<16	9.0	10
CF_4	>8	>11		12	21
SF_6	>8	14		9.4	25
C_2F_6	>8	>14		18	25
C_3H_8	>8	>20	>20	17	
C_3F_8	>8	>20		25	

[a] Data from Troe (1979).

$$\frac{\Delta E}{kT} \approx \frac{\beta_c}{1 - \sqrt{\beta_c}} \qquad (\Delta E < kT) \tag{5.32}$$

The condition $\Delta E < kT$ corresponds to collision efficiencies below 30% approximately. Typical values of ΔE are given in Table 5.1.

For example, the reaction $I + NO_2 \rightarrow INO_2$ carried out at room temperature ($RT \approx 2.5$ kJ/mol) exchanges on the average 0.8 kJ/mol with a surrounding bath gas consisting of H_2, which corresponds to a collision efficiency of about 20% only. Note that ΔE becomes considerably higher for media consisting of heavier molecules and that ΔE may exceed the thermal energy RT significantly.

Table 5.1 illustrates energy transfer is a rather efficient process. It is more efficient at lower temperatures. Molecules with many degrees of freedom are efficient in energy transfer, because now not only does energy transfer from vibrational to translational states take place, but also the more efficient conversion of vibrational energy into rotational or vibrational states of the colliding molecule occurs.

It is important to study the less efficient process of translational to vibrational energy transfer in more detail, since this is also an essential process when atoms or molecules collide with a surface. We illustrate the process with a collision between an atom and a diatomic molecule.

Figure 5.12a shows the interaction potentials of B and C in the molecule, and that of the interaction between the atom A and the molecule. The energy plot shows contours of constant energy. The dashed line represents the minimum energy path for a head-on collision between A and BC, when the latter is in the ground state. In fact, the A–BC interaction potential is given along this curve. Note that the molecule BC is compressed when A comes on the repulsive part of the A–BC potential. A collision which proceeds according to this minimum energy trajectory leaves the molecule vibrationally unexcited and the atom rebounds along the same path.

Figure 5.12b shows a trajectory where the atom A comes in at a higher energy. After the collision the molecule is vibrationally excited as inferred from the oscillation of the distance between B and C when the atom A flies away. Conservation of energy requires that A has a lower translational energy than before the collision.

The characteristic energy values for vibrational to vibrational as well as rotational to rotational energy transfer and their weak dependence on temperature relates to the phenomenon of resonance. Transfer becomes less effective when the difference in energies of the modes involved becomes larger. This follows from so-called surprisal analysis (a statistical method) and is due to decreasing cross sections when mode frequencies v become very different. According to surprisal analysis the rate constant for transitions between states labeled v and v' depends "Boltzmann-like" on the energy difference between them:

$$k(v \rightarrow v', T) = k^0(v \rightarrow v', T)e^{-(\lambda_v|E_v - E_{v'}|)/k_BT} \tag{5.33}$$

in which λ_v is a constant of the system under study. The rate constant k^0 is calculated on the assumption that rate constants are proportional to the area in phase space available on the basis of energy and momentum transferred. Relation (5.33) is derived when one optimizes the entropy $S(v)$ of the vibrational distribution with the constraint of constant energy transfer (Levine, 1981). The larger λ_v is in Eq. (5.33), the smaller the amount of energy transferred per collision.

The rate of vibrational to translational energy transfer between a colliding atom and a molecule becomes more efficient with increase of their relative translational energies. Note the contrast with the collision efficiency factor β_c, used in the rate constant (5.31), that increases when the temperature is lowered. The rate constant k has a much stronger temperature dependence than the rate of energy transfer.

The translational to vibrational energy cross section depends critically on the collision time t_c with respect to the period of the molecule vibration t_v. Consider a colinear collision of an atom A with a molecule $B(1)$-$B(2)$, where 1 and 2 number the atoms. When the collision time t_c is short compared to the vibrational period t_v during the period of interaction atom $B(2)$ does not experience any effect of the

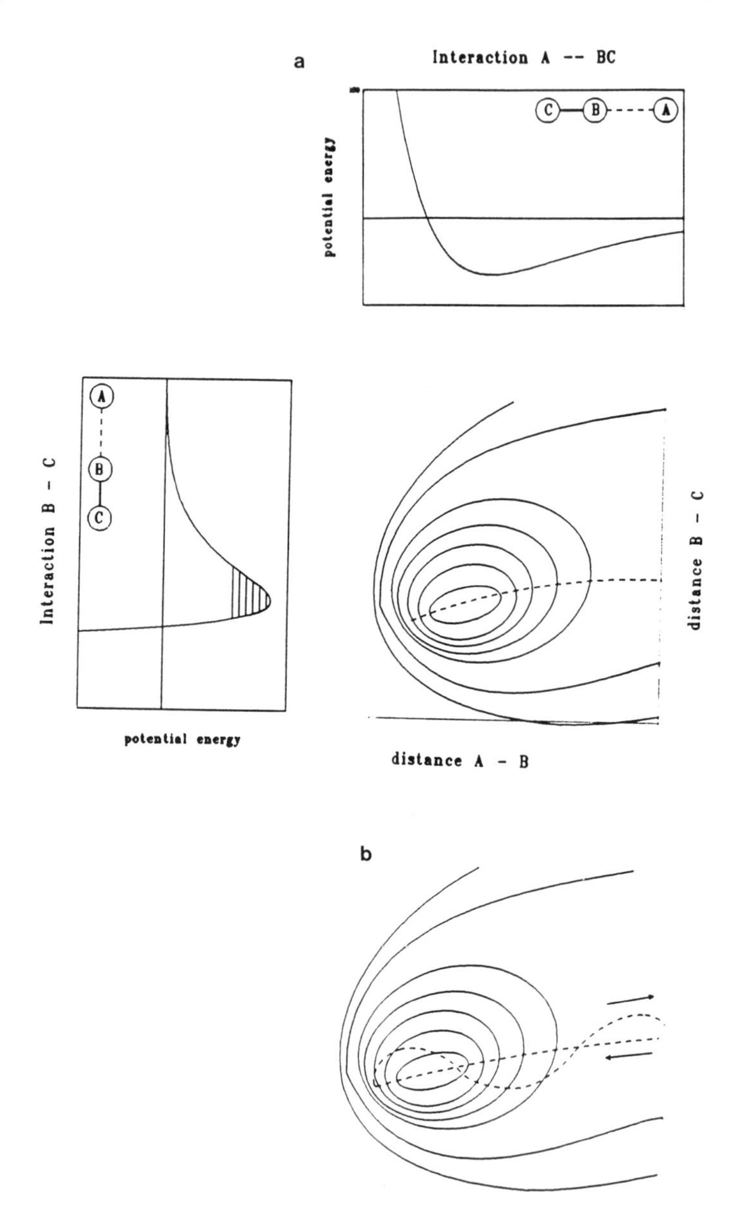

Figure 5.12. (a) Potential energy diagrams and contour maps for a head-on collision between an atom A and a molecule BC which is initially in the ground state. The trajectory in (b) corresponds to a process in which translational energy of A has been converted to vibrational energy of molecule BC.

interaction between A and $B(1)$. Energy transferred from A to $B(1)$ cannot bounce back, so that the translational energy transferred remains in the molecule as a vibrational excitation. As will be seen, the interaction is dominated by the repulsive part of the interaction potential, which has the form $\exp(-r/a)$ of length scale a. The relative velocity follows from (4.108). So the condition for efficient energy transfer becomes

$$\left(\frac{8kT}{\pi\mu}\right)^{1/2} > \frac{a}{t_v} \tag{5.34}$$

Thus at the same temperature the energy transfer cross section is larger for light atoms, because of their higher kinetic energy. This result will be very useful when thermal accommodation on surfaces is examined.

It is also of interest to consider the classical expression for energy transfer in elastic collisions. Using conservation of momentum and kinetic energy, the classical expression for energy transfer between two hard spheres (one being at rest before the collision) is

$$\frac{\Delta E}{E_0} = \frac{4\mu \cos^2 \Psi}{(1+\mu)^2}$$

$$\cos \Psi = \frac{b}{r} \tag{5.35}$$

$$\mu = \frac{M_{\text{gas}}}{M_{\text{rest}}}$$

b is the impact parameter (Figure 4.6) and r the radius of the sphere. If one averages over the collision angle Ψ or impact parameter b, the average value is

$$\alpha = <\frac{\Delta E}{E_0}> = \frac{2\mu}{(1+\mu)^2} \tag{5.36}$$

Thus, energy transfer increases the lighter the mass of the atom at rest is. The general validity of expression (5.36) is nicely illustrated by the values in Table 5.2, which shows the number of collisions with He (Z_{10}) required to deactivate a vibrationally excited molecule. A significant increase in the number of deactivating collisions required can be observed, when the molecular vibrational frequency increases. Note that small organic molecules require only a few collisions. Using a semi-classical approach, Landau and Teller derived the following expression for the probability of vibrational energy transfer

TABLE 5.2. The Number of Collisions with Helium Atoms Required to De-excite Molecules from the First Excited State to the Ground State[a]

Molecule	T (K)	Z_{10}	ν (cm^{-1})
N_2	556	2.7 10^7	2331
	761	4.3 10^6	
	1020	4.3 10^6	
	2153	1.5 10^5	
	5545	2.2 10^3	
O_2	761	1.3 10^5	1554
	1030	8.2 10^4	
	2300	3.7 10^3	
	3112	1.0 10^3	
Cl_2	300	4.3 10^4	557
Br_2	300	5.6 10^3	321
I_2	385	1.0 10^3	213
CO_2	298	6.0 10^4	672
CS_2	296	6.0 10^3	397
CH_4	303	1.5 10^4	1306
CH_3F	373	5.9 10^3	1048
CH_3Cl	298	1.0 10^3	732
$CHCl_2F$	300	5.1 10^1	274
C_3H_8	283	9	202
C_4H_{10}	300	1	

[a]From Kohlmaier (1968).

$$p_{10} = \frac{32\pi^4 m^2 \nu}{\frac{h}{2\pi} M \alpha^2} \exp\left(-\frac{4\pi^2 \nu}{\alpha \upsilon}\right) \tag{5.37}$$

in which M is the reduced mass of the oscillator, m the reduced mass of the system formed by the atom and the oscillator, ν the frequency of the oscillator, υ the relative velocity between the oscillator and the atom, and α a parameter of the potential. The exponential dependence on the frequency is also expected on the basis of surprisal analysis (5.33).

Finally, we look at the time scales on which energy exchange occurs. Suppose there is an activated reaction complex that just went through the transition state, in a medium of equilibrated molecules. The excited molecule may transfer its excess energy in collisions with medium molecules by exciting these to vibrational or rotational levels, or by transferring translational energy. In spite of the three possibilities, de-excitation measurements often show only one characteristic relaxation time. The reason is schematically indicated in Figure 5.13. Transfer of energy between rotational (R) and vibrational (V) states happens much faster than

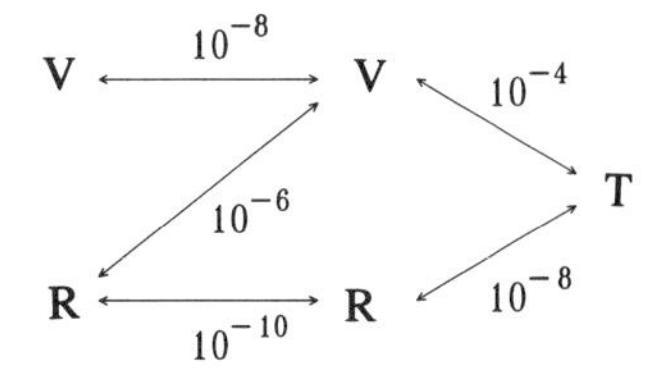

Figure 5.13. Time scale at which energy is transferred between vibrational (*V*), rotational (*R*) and translational (*T*) states. Numbers represent the product of pressure and relaxation time and are given in bar.sec. (after Flygare, 1968).

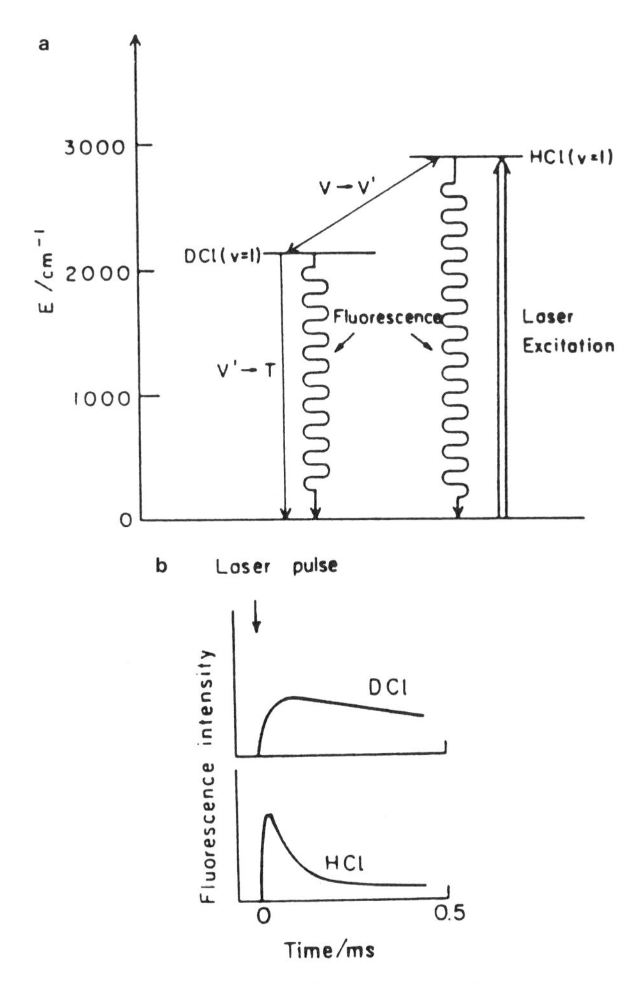

Figure 5.14. HCl in a mixture with DCl and Ar is vibrationally excited with a laser pulse. The level scheme shows the decay processes $V \rightarrow V'$ and $V' \rightarrow T$, along with fluorescence, used to determine the concentrations of HCl^* and DCl^* as shown in (b). (after Chen and Moore, 1971).

transfer from vibrational to translational (T) states. As a result, all rotational states belonging to a vibrational level will be rapidly equilibrated, while vibrational states also can be considered in thermal equilibrium. The final step in the de-excitation is the transition from the first vibrational level to rotational or translational energy, (Table 5.2). These are the slowest processes, and thus determine the overall relaxation time.

The existence of different time scales in the different modes of energy transfer is illustrated with an example. Suppose there is a mixture of HCl and DCl in an inert bath gas of argon. A short laser pulse is used to excite the HCl selectively to the first vibrational level (Figure 5.14). The DCl molecules are hardly affected by the laser pulse, at least not directly. According to Figure 5.13, the excited (in molecular dynamics jargon: vibrationally hot) HCl molecules can lose their excess energy most efficiently by exciting DCl molecules to the first vibrational level (which has a lower energy than that of HCl due to the higher mass of DCl). Loss of energy to translational states of the bath gas or rotational states of HCl and DCl is a less efficient process. The excited DCl molecules, however, have no other choice for losing their energy than through $V-R$ and $V-T$ transfer. The concentrations of HCl^* and DCl^* can be measured via fluorescence. Figure 5.14b confirms the expected behavior. The HCl^* concentration decays rapidly after the laser pulse has been given, with a concomitant increase in the concentration of DCl^*, which indeed decays at a rate that is at least two orders of magnitude slower.

5.5. THE SURFACE THERMAL ACCOMMODATION COEFFICIENT

Atoms equilibrate efficiently with the temperature of a surface when their kinetic energy is small compared to the heat of adsorption. The thermal accommodation decreases, however, with increasing energy of the atoms. The minimum occurs when the kinetic energy is on the order of the depth of the potential well for adsorption. For higher kinetic energies thermal accommodation increases again to the value for collision with a free atom of the surface material.

The sticking coefficient, i.e., the probability that a molecule impinging on a surface becomes actually adsorbed, is often smaller than unity. This implies that many molecules rebound from the surface back into the gas phase. The question arises to what extent such molecules exchange energy with the surface.

The thermal accommodation coefficient reflects the rate at which molecules colliding with a surface lose their energy. It is defined as

$$\gamma = \frac{T_r - T_g}{T_s - T_g} \tag{5.38}$$

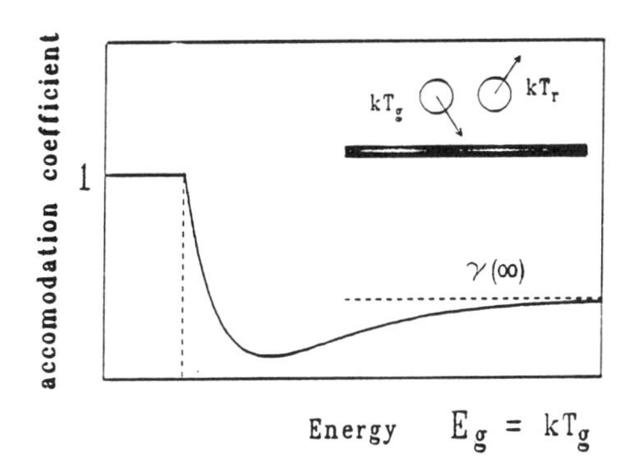

Figure 5.15. The thermal accommodation coefficient as a function of E_g.

in which T_r is the temperature of the gas molecules in the rebound stream, T_g the gas temperature of the colliding atoms and T_s the surface temperature. The qualitative dependence of the thermal accommodation coefficient on energy is sketched in Figure 5.15. The temperature of the atoms is linear in the energy (4.59), and hence

$$\gamma = \frac{\bar{\varepsilon}_r - \bar{\varepsilon}_g}{\bar{\varepsilon}_s - \bar{\varepsilon}_g} \tag{5.39}$$

Classical theory on accommodation coefficients predicts that in the limit of very high temperatures

$$\gamma(\infty) = \frac{2.4 \dfrac{M_{\text{gas}}}{M_{\text{surf}}}}{(1 + M_{\text{gas}}/M_{\text{surf}})^2} \tag{5.40}$$

where M_{surf} is the mass of an atom in the surface. This is the classical energy transfer result (5.36) but with a higher coefficient. The factor 2.4 derives from the surface geometry that limits averaging of the collision angle Ψ to a limited range. Thermal accommodation becomes most efficient if the masses of the gas molecule and the atom of the surface are the same (Table 5.3).

Accommodation coefficients on bare surfaces appear to depend only on the temperature T_g of the incident molecules, but are insensitive to surface temperature. This insensitivity to surface temperature forms the basis of the following analysis, where we show that the interaction is dominated by the repulsive part of the potential.

TABLE 5.3. Thermal Accommodation Coefficients $\gamma(\infty)$ for High Energies of the Gas Molecules

Substrate	Mass	H_2 (M = 2)	N_2 (M = 28)	Xe (M = 131)
C	12	0.29	0.5	0.18
Si	28	0.15	0.6	0.35
Fe	56	0.08	0.53	0.50
Rh	103	0.04	0.40	0.59
Pt	195	0.02	0.26	0.58

We will study the energy transfer between two particles that interact according to a Morse potential (Figure 5.16)

$$V(z) = E_0\left(e^{-2a(z-z_0)} - 2 \cdot e^{-a(z-z_0)}\right) \tag{5.41}$$

The resulting force $F(z) = -\partial V/\partial z$ becomes

$$F(z) = -2aE_0\left(e^{-2a(z-z_0)} - e^{-a(z-z_0)}\right) \tag{5.42}$$

At the turning point, defined in Figure 5.16 there is $V(z) = E_g$ (the kinetic energy of the incoming particle) and the corresponding value of z follows from

$$e^{-a(z-z_0)} = 1 + \sqrt{1 + E_g/E_0} \tag{5.43}$$

Substitution in (5.42) gives for the maximum repulsive force

$$F^{\text{rep}}_{\text{max}} \approx -2aE_0\sqrt{1 + E_g/E_0}\,(1 + \sqrt{1 + E_g/E_0}) \tag{5.44}$$

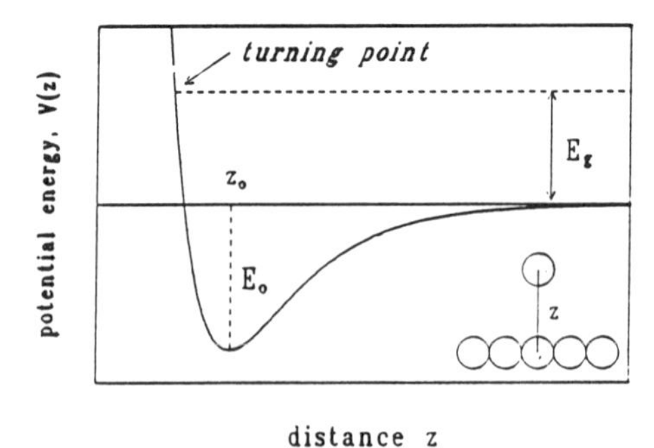

Figure 5.16. Interaction incoming particle with kinetic energy E_g with the interaction potential.

From (5.42) the maximum attractive force can be readily derived

$$F_{\max}^{\text{attr}} = +\frac{1}{2} a E_0 \tag{5.45}$$

so that

$$\left| \frac{F_{\max}^{\text{rep}}}{F_{\max}^{\text{attr}}} \right| \approx 4 \sqrt{1 + E_g/E_0} \, (1 + \sqrt{1 + E_g/E_0}\,) \tag{5.46}$$

At low energy this ratio is 8, and it increases with energy. So, as mentioned earlier, it is clearly the repulsive part of the potential that provides the most significant interaction.

For small E_g the repulsive forces are independent of E_g/E_0 and so is energy transfer ΔE to a cold surface. As long as $E_g < E_0$, the gas atom loses all its energy. The gas atom is slowed down in the surface potential well and because of its long residence time it loses its energy to the surface: the accommodation coefficient γ has become 1. Section 6.2 will discuss the dynamic correction that arises when thermal accommodation is slow.

When the energy of the incoming atom increases, it will not become fully accommodated in one collision, but it loses only part of its energy. The accommodation coefficient will have a minimum as a function of energy, because at infinitely high energies it becomes equal to the energy transfer coefficient predicted for a collision with a free surface atom.

The previous section noted that a lower kinetic energy results in a longer collision time t_c, which decreases the energy transfer. The minimum in surface accommodation coefficient is found for energies $E_0 \approx E_g \approx kT$. It can be concluded that only for physical adsorption and high temperatures does thermal accommodation at the surface becomes a slow process, but that in other situations rapid thermal equilibration applies.

5.6. MARCUS THEORY OF PROTON AND ELECTRON TRANSFER IN LIQUIDS

For reactions in solution the total interaction energy is the sum of the interaction energy between the molecules in vacuum and the energy due to the reorganization of the solvent molecules. In polar media, molecules may dissociate in ions because the ions can be hydrated by solvent molecules. Reactions that occur via a polar transition state proceed faster in a polar medium. If such reactions are fast (electron transfer), the reorganization of solvent molecules may be slow in comparison to the changing charge distribution between the reacting molecules and as a result the stabilization by the medium may decrease. The Polanyi expressions relate the activation energy to the overall free energy difference for the reaction.

In a liquid the molecules or ions interact strongly with each other. A liquid can be considered disordered on a long distance scale with respect to molecule–molecule distances. However, on a short-range, when one considers the interaction between nearest or next nearest neighbor atoms or molecules, a liquid shows order. This derives from the interaction potential between the particles.

When two particles in a liquid interact, the rate expression becomes

$$k_{\text{liq}} = k_{\text{gas}} \int d^3r\, p(r)\, e^{-W(r)/kT} \tag{5.47}$$

Here $p(r)$ is the radial distribution function (i.e., the probability that the two particles are at a distance r) in the solution and $W(r)$ is their interaction potential. Additional effects have to be taken into account when reactions involving charge transfer occur. In section (5.1), it was mentioned that polar transition states become stabilized in polar liquids, i.e., liquids with a large dielectric constant. This derives from the reorientation of the liquid molecules in the field generated by the dipole of the polar transition state. The cost to separate the charges, produced upon ion formation, has become reduced. The equation for the electrostatic interaction between two charges is

$$E_{\text{int}} = \frac{q_1 q_2}{\varepsilon r_{12}} \tag{5.48}$$

The larger the dielectric constant ε, the less the energy cost of charge separation. In a solvent the finite ion concentration leads to a screened overall interaction potential $V(r)$ between the solvated ions.

An ion dissolved in a medium will attract other ions of opposite charge. According to the Debije–Hückel theory, the concentration of the counter ion decreases exponentially with distance, the screening constant is given by $1/\kappa$, with

$$\kappa^2 = \frac{e^2}{\varepsilon T} \sum_{\alpha} Z^2 \frac{N_\alpha}{V}$$

The interaction potential becomes

$$V(r) = \frac{Ze\, e^{-\kappa R}}{4\pi\varepsilon R} \tag{5.49}$$

Z is the valence state of the ion. In a polar medium with a high dielectric constant, κ^{-1} becomes very large and the positive and negative ions can be considered to move rather independently. This solvation of cations and anions by polar solvent molecules is responsible for dissociation reactions as

$$RH \rightleftarrows H^{+} + R^{-} \tag{5.50}$$

or the dissociation of salts. In a vacuum the energy cost to separate charges is such that ions of opposing charge remain associated. Only when the ions stabilize by interaction with a medium charge, separation becomes possible. The process (5.50) is driven by the hydration energy of the product ions by solvent molecules. The acidity of a molecule generally depends on the medium as well as on concentrations in a medium.

Rudolph Marcus realized that when a charged particle moves fast through a polarizable medium (such as an electron in charge transfer processes important in electrochemistry, or H^+ in acid–base catalysis) the movement of liquid molecules may be slow compared to that of the reacting particle. According to the Debije–Hückel theory or microscopic liquid theory, the solvent molecules or ions rearrange such that there is maximum screening of charges. When the charge moves and the solvent atom positions do not follow this motion, the result is an increase in energy. Formulas like (5.49) no longer apply.

For electron transfer between identical complexes such as

$$Fe^{2+}\,(L)_n + Fe^{3+}\,(L)_n \rightarrow Fe^{3+}\,(L)_n + Fe^{2+}\,(L)_n \tag{5.51}$$

the potential energy curve as a function of reaction coordinate can be schematically represented as in Figure 5.17. Λ is the activation energy. In (5.51) the system behaves thermodynamically neutral, as in example (5.50). In cases where there is a change in free energy ΔG, the expression for $\Delta G^{\#}$ becomes

$$\Delta G^{\#} = W(r) + \Lambda\left(1 + \frac{\Delta G}{4\Lambda}\right)^{2} \tag{5.52}$$

Expression (5.52) is derived if one assumes the two potential energy curves in Figure 5.17 to be parabolic and ΔG to be small compared to Λ.

When ΔG is varied by variation of ligands or substituents, (5.52) predicts a relationship between $\Delta G^{\#}$ and ΔG

$$\Delta G^{\#} \approx W(r) + \Lambda + \frac{\partial \Delta G^{\#}}{\partial \Delta G}\,\delta \Delta G$$

$$\alpha = \frac{\partial \Delta G^{\#}}{\partial \Delta G} = \frac{1}{2}\left(1 + \frac{\Delta G}{4\Lambda}\right) \tag{5.53}$$

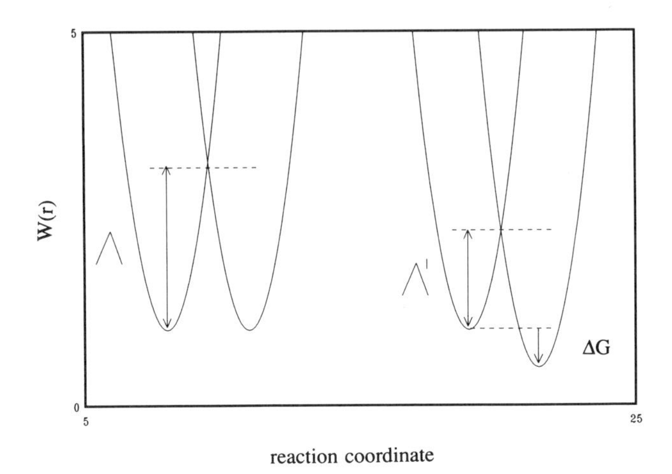

Figure 5.17. Potential energy versus reaction coordinate for a electron transfer between identical complexes.

Note that (5.53) is a linear relationship between free energy of activation $\Delta G^{\#}$ and the free energy change in overall reaction $\delta\Delta G$. This relationship is called the Polanyi–Brønsted relation. When the reaction studied is exothermic, α is smaller than 1/2. For an endothermic reaction it is larger than 1/2.

The contribution to the dielectric constant consists of a contribution due to the rearrangement of atom positions as well as a contribution due to the rearrangement of electrons. The time scale of the first is of the order of the inverse of the vibrational frequencies, the time scale of the latter is that of electronic excitation. The latter is a factor of a thousand faster than atomic rearrangement. Therefore, when an electron or a proton is transferred, rearrangement of electrons in solvent molecules will always occur.

The two effects, electronic or molecular rearrangement, can be discriminated by measuring the dielectric constant at different frequencies using electromagnetic radiation. When the frequency is low, the responses of molecular as well as electronic motion are measured. At high frequency, however, only the response due to adaptation of electronic motion is found.

In the case of an electron being transferred, Marcus (1960) deduced the following relation for Λ

$$\Lambda = m^2 \Delta e^2 \left(\frac{1}{2a_1} + \frac{1}{2a_2} - \frac{1}{a_1 + a_2} \right) \left(\frac{1}{\varepsilon(\infty)} - \frac{1}{\varepsilon(0)} \right) \tag{5.54}$$

in which $m = 1/2$ for a system like (5.52). In the transition state, half of an electron is transferred.

In (5.54) we recognize the exclusively electronic response of the medium. From $\varepsilon(\infty)^{-1}$ the term due to the rearrangement of the nuclei has to be substracted. The *a*s are the radii of the complex molecules. Their solvation energy changes due to a change in charge, but no rearrangement of the solvent nuclei occurs resulting in only partial screening.

5.7. SURFACE OVERLAYERS AND INHOMOGENEITY

Lateral interactions between adsorbates lead to non-ideal mixing behavior in surface adsorbate layers. Long-range correlation effects are the consequence of short range adsorbate–adsorbate interactions. Demixing of the adsorbate overlayer into adsorbate layer islands causes surface reactions to occur at the boundaries.

5.7.1. Long-Range Coadsorbate Effects

The interaction potentials between adsorbates are short range. These interactions form the basis of long-range effects that lead to ordered phases or to segregation of adsorbates into islands of equal molecules. The consequences for reaction kinetics can be quite important as will be seen.

Surface overlayers formed by molecular or dissociative adsorption of molecules on single crystal surfaces are often found to have a high degree of ordering. Figure 5.18 illustrates this for the adsorption of carbon atoms on the (100) surface of nickel. The atoms adsorb in high coordination sites with strong covalent bonds, that cause small changes in the position of the surface atoms bonded to the carbon atom. The interaction between adatoms is usually repulsive. As a consequence, the adsorption energy usually decreases at higher surface coverages. Adsorbates prefer specific adsorption geometries with respect to each other. The overlayer structures for C on the Ni(100) face in Figure 5.18 show adsorption positions where two adatoms share only one metal surface atom between each other. Minimum metal-surface atom sharing leads to the smallest repulsive interactions. Bonding of an adatom to a surface-metal atom weakens the affinity of this metal atom to bond with an additional adatom. This causes the repulsive interaction between adatoms when they share a surface atom. Because the interaction between two C atoms adsorbed in neighboring sites sharing two surface-metal atoms is more repulsive, than between C atoms adsorbed in neighboring sites that share one surface atom, interchange of an adsorbed C atom between these two positions leads to an effective attraction. The ordered surface phase implies that a long-range correlation exists between the atom positions, notwithstanding the short range of the adatom–adatom interaction.

The situation changes when adsorbates form polar or ionic bonds with the surface. For example, oxygen atoms on a silver (110) surface form linear rows

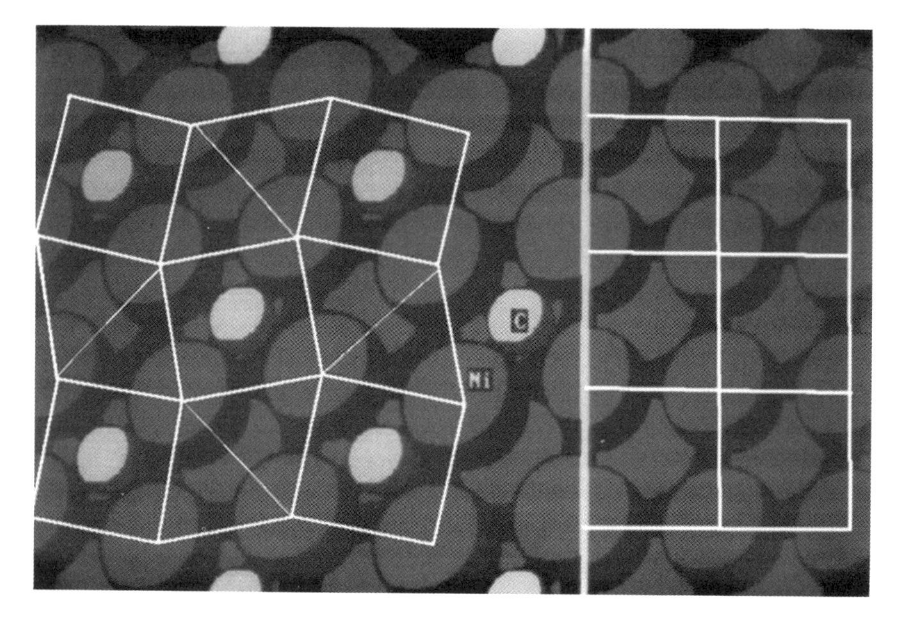

Figure 5.18. The surface structure of carbon on Ni(100) and the structure of the clean Ni(100) surface. (from Somorjai and van Hove, 1989).

perpendicular to the rows of the (110) substrate (Figure 5.19). At first this may seem unexpected, but the explanation is that the Ag–O bond has considerable polarity and that the O rows can actually be viewed as an ionic chain: $Ag^{\delta+}$-$O^{\delta-}$-$Ag^{\delta+}$-$O^{\delta-}$-$Ag^{\delta+}$, which is energetically rather favorable. At low oxygen coverage the rows are parallel and far from each other, probably as a result of repulsive $Ag^{\delta+}$–$Ag^{\delta+}$ and $O^{\delta-}$–$O^{\delta-}$ interactions between the rows.

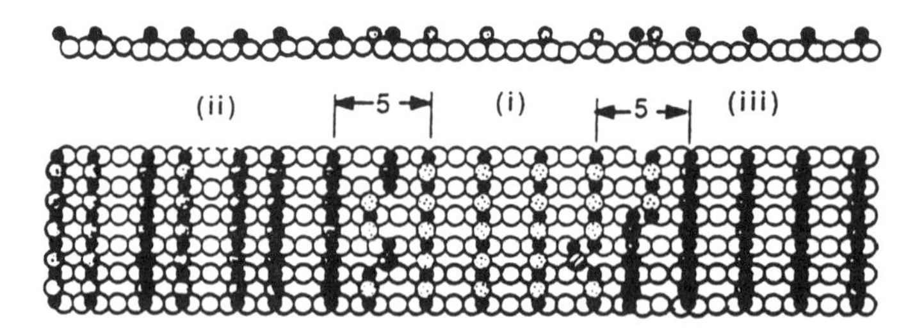

Figure 5.19. Overlayers of adsorbed oxygen atoms on the Ag(110) surface from Taniguchi *et al.*, 1992).

When dealing with practical inhomogeneous systems, composed of metal particles that expose different surfaces upon adsorption, the more reactive surfaces will become covered first and, with increasing coverage, also the less reactive surfaces. Hence, a decrease in heat of adsorption with coverage results. A decrease in heat of adsorption not only reflects repulsive interaction energies but also differences between surface sites.

A surface layer consisting of interacting adatoms can be considered as a two-dimensional non-ideal mixture. In case of adsorption of one component, the mixture exists of occupied and nonoccupied sites. When adatoms have a repulsive interaction when adsorbed on neighboring sites, the mixture is characterized by an attractive interaction energy between the vacant and the occupied site. Then a critical temperature exists such that below this temperature there is an ordering of vacant and occupied sites and above it disordering occurs. An approximate expression for the critical temperature can easily be derived within the Bragg–Williams approximation. Above the critical temperature the arrangement of the particles is considered to be random. The corresponding entropy of mixing is

$$S = \theta \ln \theta + (1 - \theta)\ln(1 - \theta) \tag{5.55}$$

The interaction energy between occupied and vacant sites is $1/2\omega\theta^2$, where ω is the interaction energy. The chemical potential follows from minimization of the free energy with respect to θ

$$\mu = \omega\theta + kT \ln\frac{\theta}{1 - \theta} \tag{5.56}$$

When the surface coverage equals 1/2, $d\mu/d\theta = 0$ which gives for the critical temperature relation $T_c = \omega/4k$.

An interaction energy of a few kilocalories per mole results in a critical temperature of a few hundred Kelvin. Below the critical temperature T_c occupied and vacant sites are ordered, which gives rise to a long-range correlation between the occupied adsorption positions. Remember, this long-range effect is based on forces that are short range and do not need to be larger than the nearest-neighbor site distances.

In reacting systems, under special conditions the composition of the surface phase may also be controlled by long-range correlation effects. Due to differences in adsorption energies the relative surface concentration of different components is a function of temperature. Usually the surface concentration of one component, the "*mari*" dominates, but the surface composition is a function of temperature. At a transition temperature (often the maximum in the rate versus temperature curve (see Figure 2.9), the *mari* starts to change. At this transition temperature several components will be coadsorbed in comparable quantities.

An interesting example is provided by CO oxidation. Here the order–disorder phase transition is due to mixing of the CO and adsorbed O overlayers. When the interaction energy between adsorbed CO and O atoms is repulsive and larger in value than the average of the interaction energies of CO in a CO adlayer and O in an O adlayer, a critical temperature exists such that below this temperature O and CO separate into single component surface adlayer islands. A surface phase diagram can be constructed that relates the surface concentrations to the temperature of demixing. Above the critical temperature, O and CO will disorder and mix. Below the critical demixing temperature, reaction between O and CO can only occur at the boundary of the islands. The overall rate of reaction will be significantly suppressed compared to the rate computed for an ideal surface mixture and the overall reaction rate will be significantly enhanced. Above the critical temperature, reaction occurs in the homogeneous surface mixture.

Lateral interactions between adsorbates change the expressions for the surface rate constants as discussed shortly in Section 5.1. This will be illustrated for the rate expressions for molecular and associative desorption. When adsorbates interact and molecules A desorb in the presence of molecules B, the rate of desorption of molecules A is given by

$$-\frac{d\theta_{\mathrm{A}}}{dt} = k_{\mathrm{des}}\,(\theta_{\mathrm{A}},\theta_{\mathrm{B}}) \cdot \theta_{\mathrm{A}} \tag{5.57}$$

Whereas in the absence of adsorbate–adsorbate interactions k_{des} is a rate constant independent of surface composition, it becomes also a function of θ_{A} and θ_{B} in the presence of lateral interactions. This dependence will be derived using transition-state theory.

If $W_{\mathrm{A}_i^*}$ is the probability that A on site i desorbs via the activated complex A^*, then according to transition-state theory

$$-\frac{d\theta_{\mathrm{A}}}{dt} = \sum_i r_{\mathrm{des},i} = \sum_i \frac{kT}{h} W_{\mathrm{A}_i^*} \tag{5.58}$$

and $W_{\mathrm{A}_i^*}$ is given by

$$W_{\mathrm{A}_i^*} = \frac{p.f.'^{\#}_{\mathrm{A}}}{p.f.'_{\mathrm{A}}}\, e^{-(E_{\mathrm{des}} + \Delta\varepsilon_i)/kT}\, W_{\mathrm{A}_i} \tag{5.59}$$

where W_{A_i} is the probability that site i is occupied by A. Each site is characterized by an interaction energy $\Delta\varepsilon_i$; $p.f.'^{\#}_{\mathrm{A}}$ is the partition function of the desorbing molecule in the transition state, without the contribution due to the reaction coordinate and $p.f.'_{\mathrm{A}}$ is the partition function of the adsorbed molecule. In expression (5.59), the partition functions are assumed to be independent of coverage (for a discussion of the effect of lateral interactions on the molecular partition functions see the next section). E_{des} is the corresponding barrier energy for desorption of an isolated absorbed species. W_{A_i} depends on the surface coverage of A, θ_A

$$W_{A_i} = \theta_A \cdot P_{A_i} \tag{5.60a}$$

with

$$\sum_i P_{A_i} = 1 \tag{5.60b}$$

P_{A_i} is the probability that a site occupied by A has environment i. The transition-state expression for the effective rate constant of desorption can be rewritten as

$$k_{des}(\theta_A,\theta_B) = \frac{kT}{h} \cdot \frac{p.f._A'^{\#}}{p.f._A} e^{-E_{des}/kT} \sum_i P_{A_i} e^{-\Delta\varepsilon_i/kT} \tag{5.61a}$$

$$= k_{des}^{T} \cdot F_{des}(\theta_A,\theta_B) \tag{5.61b}$$

The function F contains the effect due to interactions between the absorbed species.

If one replaces the interaction between the particles by one averaged, self-consistent interaction, the probabilities P can be computed according to a quasi-chemical approximation. For the adsorption of one type of molecule, A, one defines in this approach three quantities, W_{AA}, W_{AO}, and W_{OO}, being the probabilities that two adjacent sites are occupied by two, one, or zero particles. The equilibrium condition then gives the following relation:

$$\frac{W_{AA} \cdot W_{OO}}{W_{AO}^2} = 0.25\, e^{-\Delta\varepsilon/kT} \tag{5.62}$$

in which $\Delta\varepsilon$ is the nearest-neighbor lateral interaction difference

$$\Delta\varepsilon = 1/2(E_{AA} + E_{OO}) - E_{AO} \tag{5.63}$$

For a surface with z nearest neighbors F becomes (Zhdanov, 1991b)

$$F = \left(\frac{W_{AA}\, e^{\Delta\varepsilon/kT} + W_{AO}}{W_{AA} + W_{AO}} \right)^z \tag{5.64}$$

In a way analogous to (5.61) one can also derive the correction due to lateral interactions for the rate of associative desorption

$$k_{ass}(\theta_A, \theta_B) = k_a^T \cdot F_a \tag{5.65}$$

with

$$F_a = \sum_i P_{AA_i} e^{-\Delta\varepsilon_i/kT} \tag{5.66}$$

where P_{AA_i} is the probability that a site with a pair of recombining A atoms has the environment i. The presence of lateral interactions may result in significantly suppressed rates of reactions, compared to predictions of transition state theory for the non interacting case. Such suppressed rates are often found at catalytic conditions. Whereas the initial sticking coefficient (the sticking coefficient at low coverage) of CO on Rh is 0.5, this sticking coefficient reduces to 10^{-4} in the CO oxidation reaction at low temperature. Similarly, the sticking coefficient for dissociative oxygen adsorption on silver has been found to decrease from 10^{-2} to 10^{-7}.

Numerical procedures based on Monte Carlo techniques have been developed to solve expressions of type Eq. (5.58) rigorously, without having to assume averaging assumptions as done in the Bragg–Williams or quasi-chemical approximations.

5.7.2. Short-Range Coadsorbate Interaction

Coadsorption suppresses surface reactions that require sites consisting of several metal atoms by blocking. The close proximity of coadsorbate molecules may lead to steric hindrance effects, reducing translational or rotational motion. Short-range lateral effects may enhance as well as decrease the adsorbate-surface interaction strength. Enantiomeric selectivity as well as selective oxidation are examples of reactions that are sensitive to the composition of the surface adlayer. In the former case the selectivity is enhanced due to steric interactions. In the latter case the chemical nature of adsorbed surface atoms changes with surface adatom concentration.

The previous section analyzed consequences of lateral interactions on the statistical distribution of adsorbates over the adsorption sites. This section will discuss changes in the transition-state rate constant k_T due to changed pre-exponents and activation energies.

Coadsorbed species can interact in many different ways, when adsorbed at close distance from each other. Chapter 4 mentioned the decreased pre-exponents for rate constants of desorption of adatoms adsorbed at high coverage. The pre-exponent decreases with coverage, because of the decreased mobility of the atoms in the transition state due to site blocking by the other adatoms. When a surface is densely packed with an adsorbed layer of bulky molecules, the desorption rate of molecules may also become suppressed for molecules that desorb with a loose transition state. A loose transition state implies nearly free rotation in the transition state. This rotation may become hindered when bulky molecules are coadsorbed. Such a steric effect is exploited chemically in enantioselective catalysis.

Hydrogenation of a keto-group may generate an optically active carbon center (indicated by an asterisk)

$$
\begin{array}{lcccc}
 & R & & & R \\
 & | & & & | \\
H_2 + & C{=}O & \rightarrow & H- & C^{*}-OH \\
 & | & & & | \\
 & R^1 & & & R^1
\end{array}
$$

When a ketone is hydrogenated by a transition-metal catalyst, covered by one of the optically active enantiomers of a coadsorbate molecule (an experimental example is the adsorption of one of the enantiomeric forms of tartaric acid to nickel), the interaction between the coadsorbed optically active enantiomer and keton results in the preferential formation of one of the possible enantiomers of the alcohol. The coadsorbate plays a very similar role as a bulky ligand of a homogeneous catalyst, which influences the selectivity of a reaction by direct steric interactions. Coadsorbates may also alter the interaction of adsorbates by changing the interaction energies.

An example for such a change in heat of adsorption is provided by the effect of coadsorbed NO on the heat of adsorption of CO on Pt. The rate of desorption of CO becomes suppressed by coadsorbed NO, because of the electrostatic interaction between NO_{ads} and CO_{ads}. In the chemisorbed state of a molecule, electron transfer between metal and adsorbate occurs. The electron affinity of NO is higher than that of CO and NO becomes negatively charged. NO binds with a higher adsorption energy than CO. CO however has a low electron affinity and develops positive charge because of electron donation to the metal. When adsorbed next to each other, the two opposite charges will attract (Figure 5.20). As a consequence the desorption energy of CO has increased.

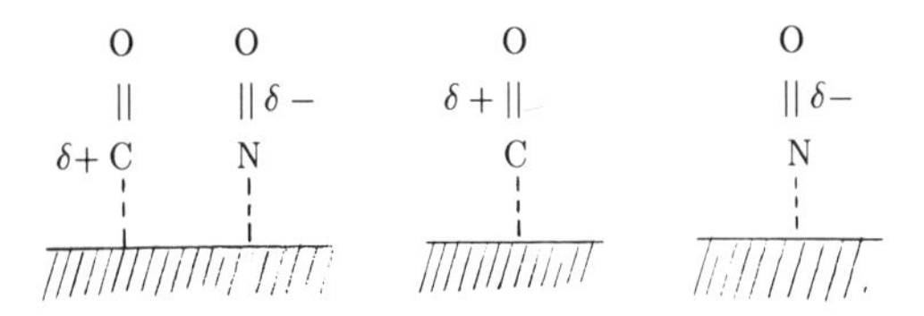

Figure 5.20. Coadsorption of NO and CO.

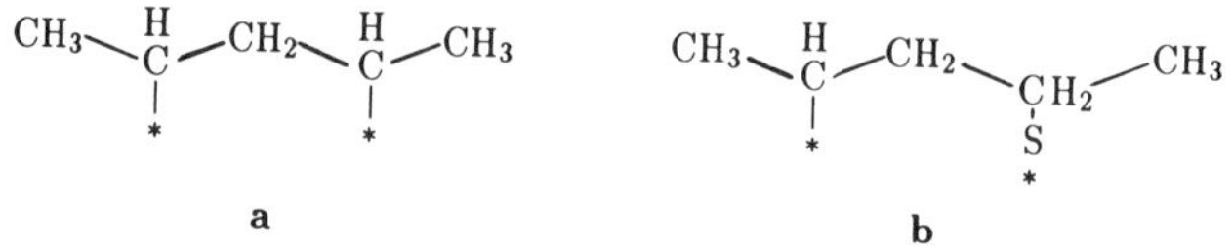

$CH_3-\underset{*}{\overset{H}{C}}-CH_2-\underset{*}{\overset{H}{C}}-CH_3$ $\quad$ $CH_3-\underset{*}{\overset{H}{C}}-CH_2-\underset{\underset{*}{S}}{CH_2}-CH_3$

a $\quad$ b

Figure 5.21. Primary ensemble effect; (a) multimolecular atom adsorption to a metal surface; (b) decrease in heat of adsorption by site blocking. S indicates an adsorbed sulfur atom, (schematic; * denotes a site consisting of one or more surface metal atoms).

Usually the presence of coadsorbates will decrease the interaction strength of admolecules with surface atoms. This may occur for several reasons. The deactivation of surface atom reactivity due to adatom adsorption is an example discussed earlier in this section.

A large molecule, e.g., a hydrocarbon, may attach to a metal surface with several bonds. This is schematically illustrated in Figure 5.21. In this figure dissociative adsorption of pentane occurs by attachment of two of its carbon atoms to the metal surface. This mode of adsorption becomes suppressed when sulfur or carbon atoms are coadsorbed. The S or C atoms decrease the probability for a surface to have neighboring-reactive sites. As a result, the probability for pentane to form multiple bonds with the metal surface becomes suppressed. It will only adsorb with one of its carbon atoms attached to the surface. As a result of this site blocking effect, the heat of adsorption of pentane decreases and the probability for dissociative adsorption decreases as well.

The presence of inert coadsorbed atoms also suppresses the rate of dissociation of a molecular bond. This will be illustrated for dissociation of a molecule as CO. Dissociation occurs by two consecutive steps:

$$CO_{gas} + * \rightarrow CO_{ads}$$

$$CO_{ads} + * \rightarrow C_{ads} + O_{ads}$$

Whereas an adsorbing CO molecule requires one vacant surface site, the two adatoms generated upon dissociation require two empty neighboring-surface sites (for a more extensive discussion of surface-dissociation reaction, refer to the next chapter). Again, site blocking by coadsorption of inert adatoms will decrease the probability for neighboring-vacant sites, and suppress the dissociation probability of an adsorbed molecule.

So far the vacant surface site has been indicated as a simple *. On an atomic level the adsorption site is an ensemble of surface atoms, as shown in Figure 5.18 for adsorbed carbon atoms: C coordinates to a site of four surface-metal atoms. In general, adsorbed molecules or atoms can adsorb to sites with different coordination numbers. Atoms usually prefer high coordination. Therefore, coadsorption of

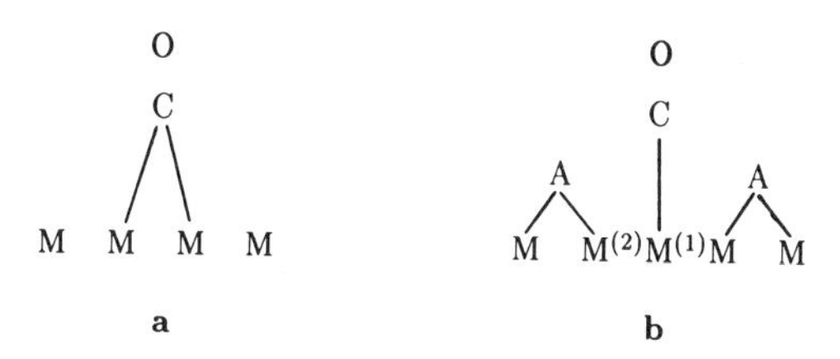

Figure 5.22. Secondary ensemble effect; (a) twofold coordination; (b) onefold coordination, adsorbate A blocks high coordination sites (schematic).

adatoms will decrease the possibility for other atoms to adsorb to high coordination sites on a surface. For adsorbates that can adsorb in different coordinations this will favor adsorption to low coordination sites. This so-called secondary ensemble effect is illustrated for CO in Figure 5.22.

The secondary ensemble effect also changes the heats of adsorption, again due to site blocking. The chemical affinity of a surface atom not in contact with a coadsorbate atom (atom 1 in Figure 5.22b) has been considered to be unaltered. Large changes in chemical affinity occur for surface-metal atoms attached to an adsorbing atom (atom 2 in Figure 5.22b). Coordination of this surface atom to an atom decreases the affinity of that surface atom to form additional bonds with more atoms. This effect was discussed earlier in this section. There it gave rise to repulsive lateral interaction between surface adatoms. The bond order conservation principle postulates that the total valence of an atom is constant and is distributed over the bonds that an atom forms with its neighbors. As a consequence, the more neighbors an atom has, the lower the bond strength per individual bond.

Bond order conservation is also the chemical origin of the changes in position of surface atoms in contact with adatoms. The metal–metal bonds weaken due to coadsorbed atoms. The same effect decreases the bondstrength of adatoms that share bonds with the same metal atom. Because of this, a surface may have completely different reactivities at low and at high surface coverage. An interesting example we discussed before is provided by the selective oxidation of ethylene on silver, (Section 3.1.2.4). Whereas ethylene is mainly oxidized to CO_2 and H_2O by a silver surface with a low coverage of oxygen, selective reaction to the epoxide occurs on oxidic silver surfaces with a high surface and subsurface content of atomic oxygen. The bond strength of oxygen adatoms to silver decreases sharply on locations of a high oxygen concentration.

Coadsorbate-induced reactivity changes are often quite important to the performance of a catalyst. The presence of a surface adatom layer is one important reason for differences in reactivity found in catalytic studies performed at conditions typical of a vacuum, (e.g., 10^{-9} Torr) compared to the conditions used in practical catalysis, typically 10^3 Torr. At low pressure the low rate of dissociative adsorption may inhibit formation of a surface layer. Also surface layers may become unstable under vacuum conditions. An example of such an unstable overlayer is the partially hydrogenated carbon layer found on metals in hydrocarbon conversion reactions. Hydrogen will evolve at low pressures and a non-reactive carbon residue is formed. Because of the weakened bond affinity of surface metal atoms by adatoms, the sticking coefficients for dissociation of molecules become reduced at high pressures.

These are the three main reasons for the existence of "pressure gap" differences as sometimes found for catalytic reactions. Nevertheless, favorable cases exist in which rate constants of elementary surface reactions that have been obtained under vacuum conditions can be used to predict overall rates of a catalytic reaction that are valid in a wide range of conditions. Examples are discussed in the next chapter.

MICROSCOPIC THEORY OF HETEROGENEOUS CATALYSIS

The most important elementary steps of the catalytic reaction are adsorption, dissociation, association, and desorption. The event is essentially a cycle in which the catalytically active sites are regenerated. Adsorption proceeds usually with or without a small activation energy. Dissociation is often activated and requires an ensemble of several surface atoms. The activation energy for dissociation may be lowered by electron back donation into antibonding orbitals of the adsorbate. Association reactions are favored by factors opposite of those favoring dissociation. The activation energy for desorption is usually comparable to the heat of adsorption. According to Sabatier's principle, an optimum exists in the interaction energy between the adsorbate and the catalytic surface that corresponds to a maximum in the rate of the catalytic reaction. The rate-limiting steps of the reaction cycle are different below or above the optimum interaction energy. Competitive adsorption of reactant and/or product molecules often controls the selectivity and activity of a catalyst.

According to Berzelius' classic definition, a catalyst is not consumed during a reaction. Its action is based on the formation of reaction intermediates between catalyst and reagent molecules. In subsequent reaction steps, the catalyst is regenerated and product molecules evolve. The overall reaction of reagent molecules to product molecules consists of a sequence of elementary reaction steps that together form a cycle. Empty sites on the surface become occupied by adsorbed reactants and intermediates of the reaction, and are regenerated when the products of the reaction desorb.

This self-regenerating cycle of elementary steps, as a catalytic reaction is defined in Chapter 2, exemplifies the ideal situation for a catalyst of infinite life. However, in practice a catalyst may gradually deactivate under reaction conditions. Non-selective side reactions may produce residue molecules that accumulate on the surface. Coke formation is a good example. In addition, the structure of the heterogeneous catalyst may change, due to solid state reactions occurring parallel to the desired chemical conversions. For instance, the small, initially well-dispersed particles that form the catalytically active component may sinter during reaction. In homogeneous catalysis non-selective reactions may affect the ligands of the active complex. Raw materials used as reactants often contain reactive impurities that block the active surface of the catalyst. In practice, deactivation is an important variable, affecting not only the activity but also the selectivity of a catalyst.

In principle, a catalyst can affect reactions in two ways. First, reactions become possible at relatively low temperatures and pressures, implying an increase in chemical activity. Second, and often as important, the catalyst may increase the selectivity of a reaction when it enhances the formation of the desired product and suppresses other products. Whereas the increase in activity is usually due to a decrease in overall activation energy of the reaction, changes in selectivity must be due to a favorable activation energy for the desired reaction, and a less favorable reaction path for other reactions.

Most catalytic reactions consist of at least four elementary reaction steps: adsorption, dissociation, association, and desorption. This is illustrated for the oxidation of CO

$$
\begin{aligned}
CO + * &\leftrightarrows CO_{ads} && (a)\\
O_2 + * &\leftrightarrows O_{2,ads} && (b)\\
O_{2,ads} + * &\leftrightarrows 2O_{ads} && (c)\\
CO_{ads} + O_{ads} &\leftrightarrows CO_{2,ads} + * && (d)\\
CO_{2,ads} &\leftrightarrows CO_2 + * && (e)
\end{aligned}
\qquad (6.1)
$$

Often the dissociation of the oxygen molecule in reaction step (*c*) occurs rapidly such that adsorbed O atoms are equilibrated with gas phase oxygen. Examples will be provided further on in this chapter.

In the absence of surface diffusion, the distribution of adsorbed species would be random. As mentioned in Section 5.7.1, island formation and adsorbate ordering processes occur under reaction conditions where the rate of diffusion of adsorbates over the surface is high. Because diffusion is usually not a rate-limiting step in a catalytic reaction sequence, it is commonly not considered explicitly in discussions of reaction mechanisms.

At present, the chemical factors controlling the activation energy and entropy of reaction steps on surfaces are extensively explored and a consistent theory starts to emerge. Kinetic studies of surface reactivity allow for the following conclusions: 1. Molecular adsorption onto a surface is usually not activated. The initial sticking probability for the adsorption of diatomic molecules on clean surfaces is mostly on the order of 10^{-2}–1; and 2. Dissociative adsorption is typically associated with much lower sticking coefficients, often 10^{-4}–10^{-8}. The reasons are twofold. Changing from the molecular to the dissociated state requires stretching of the molecular bond, which costs energy. There will be an activation barrier which, at a typical reaction temperature of 500–600 K, would reduce the sticking coefficient by four orders of magnitude. The activation energy for dissociation of the molecule in the adsorbed state is much less than in the gas phase (CO: 1076 kJ/mol in the gas phase, about 150 kJ/mol on a surface), owing to the formation of strong bonds with the surface. A low activation energy for dissociation is achieved when the reaction path is such that the dissociating atoms are close to the surface. This implies that the transition state is tight. The loss of translational and rotational freedom may contribute an additional factor of 10^{-4} to the sticking coefficient for dissociative adsorption of diatomic molecules. Thus, dissociative adsorption that is not activated with respect to the gas phase, but proceeds via a tight transition state, is expected to have a sticking coefficient on the order of 10^{-4}, whereas activated dissociation typically has sticking coefficients within the range of 10^{-4}–10^{-8}, or lower.

Because the activation energies for dissociation are low in general, it has long been thought that dissociative adsorption would not be the rate-limiting step of the catalytic reaction chain. However, as indicated above, the change in rotational or vibrational degrees of freedom in the dissociative adsorption process may result in an extremely low preexponential factor for the rate constant, which compensates for the low activation energy and reduces the reaction rate.

In later sections of this chapter the microscopic basis of the sticking coefficient will be returned to when the main aspects of chemical bonding on surfaces and the implications for the overall kinetics of a catalytic reaction are examined.

6.1. PREDICTION OF THE OVERALL RATE OF A CATALYTIC REACTION

The overall rate of a reaction can be computed once the rates of the elementary steps are known. These can be deduced from model experiments or by theoretical calculations. The assumption is often made that coverage-independent kinetic parameters or the neglect of lateral interactions may lead to sticking coefficients and preexponentials of rate constants that are too high.

One of the objectives of this book is to demonstrate how the overall reaction rate of a catalytic reaction can be predicted once the rate constants and equilibrium constants of the elementary reaction steps are known. Such data can be obtained from model experiments, often in surface science, or from theoretical calculations. The ammonia synthesis was among the first catalytic reactions for which the rate was predicted under high pressure conditions. This was a remarkable success, as the calculation involved an extrapolation of data obtained under vacuum over a pressure interval of ten orders of magnitude. Here the effective rate constant of a less complex reaction is derived, the oxidation of carbon monoxide on platinum. We will use the reaction mechanism of Scheme (6.1) and label rate and equilibrium constants according to the number of the elementary steps in (6.1).

Assuming that step 3, the formation of adsorbed CO_2 from adsorbed CO and O on the surface, is rate limiting, the overall rate of reaction becomes

$$r = V\frac{d[\mathrm{CO_2}]}{dt} = N\,k_3\,\theta_{\mathrm{CO}}\,\theta_{\mathrm{O}} \tag{6.2}$$

The coverages of adsorbed O, CO, and CO_2 follow from the Langmuir isotherms

$$\begin{aligned} \theta_{\mathrm{CO}} &= K_1[\mathrm{CO}]\theta_* \\ \theta_{\mathrm{O}} &= K_2^{1/2}[\mathrm{O_2}]^{1/2}\theta_* \\ \theta_{\mathrm{CO_2}} &= K_4[\mathrm{CO_2}]\theta_* \quad (K_4 = k_{\mathrm{ads}}/k_{\mathrm{des}}) \end{aligned} \tag{6.3}$$

while the fraction of unoccupied sites follows from the condition

$$\theta_* + \theta_{\mathrm{CO}} + \theta_{\mathrm{O}} + \theta_{\mathrm{CO_2}} = 1 \rightarrow \tag{6.4}$$

$$\theta_* = \frac{1}{1 + K_1[\mathrm{CO}] + K_2^{1/2}[\mathrm{O_2}]^{1/2} + K_4[\mathrm{CO_2}]} \tag{6.5}$$

and hence the complete expression for the rate becomes

$$r = V\frac{d[\mathrm{CO_2}]}{dt} = \frac{N\,k_3\,K_1\,K_2^{1/2}\,[\mathrm{CO}][\mathrm{O_2}]^{1/2}}{(1 + K_1[\mathrm{CO}] + K_2^{1/2}[\mathrm{O_2}]^{1/2} + K_4[\mathrm{CO_2}])^2} \tag{6.6}$$

The rate will be expressed per catalytically active site on the surface. This quantity is usually referred to as a turnover frequency

$$TOF = \frac{r}{N} = \frac{V}{N}\frac{d[CO_2]}{dt} \tag{6.7}$$

Assuming that CO is the majority reacting intermediate, as will be the case at relatively low temperatures, (6.6) reduces to

$$TOF = \frac{k_3 K_2^{1/2}}{K_1}\frac{[O_2]^{1/2}}{[CO]} \tag{6.8}$$

for the CO oxidation on platinum at relatively low temperatures. At high temperatures, however, the surface coverages of both CO and O become low, and the rate becomes

$$TOF = \frac{V}{N}\frac{d[CO_2]}{dt} = k_3 K_1 K_2^{1/2} [CO][O_2]^{1/2} \tag{6.9}$$

In order to calculate the turnover frequency, it is necessary to calculate the equilibrium constants for adsorption of CO and oxygen. The equilibrium constant equals

$$K = \frac{k_{ads}}{k_{des}} \tag{6.10}$$

The rate constants for adsorption may, according to (4.114), (4.182), and (4.183), be written as

$$k_{ads} = s_o \frac{1}{4}\bar{u}\,\pi d^2 = s_o \frac{1}{4}\left(\frac{8RT}{\pi M}\right)^{1/2} \pi d^2 \tag{6.11}$$

where πd^2 is the area of an active site (we take 1.4×10^{-15} cm^2). Using a sticking coefficient for oxygen of 0.02, the rate of adsorption is found to be $k_{ads} = 2.7 \times 10^{-13}$ cm^3/s at 450 K. As the sticking coefficient for CO at low coverage is about unity, the rate of adsorption becomes equal to $k_1^f = 1.5 \times 10^{-11}$ cm^3/s, at 450 K.

The rate of desorption is written as

$$k_{des} = \nu\, e^{-E_{des}/RT} \tag{6.12}$$

Typical values for the rate parameters have been given in Table 4.8. For CO desorption we use a preexponent of 10^{14}/s and an activation energy of desorption of 145 kJ/mol. For desorption of oxygen we use $\nu = 10^{17}$/s, and $E_{des} = 290$ kJ/mol (i.e., 145 kJ/mol per O atom). With these values the following equilibrium constants are

$$K_1 = K_{\text{eq}}^{\text{CO}} = 1.5 \cdot 10^{-25}\, e^{17.5 \cdot 10^3/T}\ (\text{cm}^3) \tag{6.13}$$

and

$$K_2 = K_{\text{eq}}^{\text{O}_2} = 2.7 \cdot 10^{-30}\, e^{17.5 \cdot 10^3/T}\ (\text{cm}^3) \tag{6.14}$$

The rate of the elementary reaction between O_{ads} and CO_{ads} to CO_2 can be determined separately by coadsorbing CO and O_2 on a metal surface and studying the rate of CO_2 formation in the gas phase using temperature programmed desorption spectroscopy. According to Zhdanov (1991b), typical values of the preexponential and activation energy are 10^5/s and 45 kJ/mol for the reaction on platinum

$$k_3 = 10^5\, e^{-5.4 \cdot 10^3/T} \tag{6.15}$$

Substitution of the values in the rate expression (6.8) yields the turnover frequency of the CO oxidation at 450 K

$$TOF - \frac{V}{N}\frac{d[\text{CO}_2]}{dt} = 1.1 \cdot 10^{15} e^{-14.15 \cdot 10^3/T} \frac{[\text{O}_2]^{1/2}}{[\text{CO}]}\, s^{-1} \tag{6.16}$$

As adsorption proceeds without an activation barrier, the activation energy of the reaction is essentially equal to the heat of desorption of CO, expressing that desorption of CO is necessary in order to make space for the dissociative adsorption of oxygen. The preexponent of the reaction is typical for reactions with some change in degrees of freedom in the transition state. The turnover frequency is low, due to the strong adsorption of CO, which blocks sites where oxygen should adsorb and dissociate. The rate increases with temperature, because the equilibrium constant of CO becomes smaller implying that more sites become available for the oxygen.

At high temperatures (above 550 K) the following expression can be obtained for the turnover frequency

$$TOF = \frac{V}{N}\frac{d[\text{CO}_2]}{dt} = 2.5 \cdot 10^{-35}\, e^{20.85 \cdot 10^3/T}\, [\text{CO}][\text{O}_2]^{1/2}\, s^{-1} \tag{6.17}$$

Notice that the overall activation energy is negative, implying that the rate goes down with increasing temperature. The situation is similar as discussed in connection with Figure 2.9: At low temperatures the rate of reaction, as given by (6.8), increases with temperature. As the surface is predominantly covered by CO, dissociative adsorption of oxygen limits the rate. At higher temperatures the rate reaches a maximum; here the full expression (6.6) applies. At increasingly higher temperatures, the surface becomes more and more empty and the association between CO and O atoms determines the rate, which decreases with temperature

TABLE 6.1. Preexponential Factors, Activation Energies at Low Coverages and Lateral Interaction Energies between Nearest Neighbor Atoms[a]

Surface	ν (s^{-1})	E_{act} (kJ/mol)	$\varepsilon_{CO\text{-}CO}$ (kJ/mol)	$\varepsilon_{O\text{-}O}$ (kJ/mol)	$\varepsilon_{CO\text{-}O}$ (kJ/mol)
Ir(110)	10^{13}	155	9.2	15.1	7.1
Ir(111)	10^{13}	130	2.9	3.4	2.1
Pt(111)	2×10^{14}	100	11.8	19.3	10.0

[a]From Zhdanov (1991).

according to (6.17). Thus, the majority reacting intermediate changes from CO at low temperature to the unoccupied sites at high temperature; the change is accompanied by altered reaction orders in the pressures of CO and O_2.

The preceding discussion serves to illustrate how the overall reaction rates can be predicted based on elementary surface reaction rate constants measured in model experiments. This section ends with a few remarks. First, no interaction between the adsorbed species was included in the discussion above. Lateral interactions certainly play an important role, particularly at intermediate temperatures where oxygen atoms tend to form islands. This implies that the reaction proceeds predominantly at the boundaries of the oxygen islands, which reduces the reaction rate considerably. Table 6.1 lists a number of preexponentials, activation energies, and interaction energies for the CO oxidation on different substrates. The preexponents of the rate constants for the surface reaction between O_{ads} and CO_{ads} are as expected. If the lateral interactions in the system were ignored, there would be significantly lower values of the kinetic parameters. According to Zhdanov (1991b), the preexponential value drops to 10^5 s^{-1} and the activation energy to 45 kJ/mol. This illustrates the consequence of long-range, order–disorder effects on the effective rate constants.

Second, the coverage dependence of the sticking coefficients for the adsorption of CO and O_2 was ignored. As will be explained in the next section, sticking coefficients may become significantly smaller at higher coverages. Assuming that the sticking coefficients of CO and O_2 depend similarly on coverage, the effect cancels more or less at low temperature, as evident from (6.8). At high temperatures, all coverages are low and the sticking coefficients are essentially equal to s_0. At intermediate temperatures, however, the coverage-dependent effects on the sticking coefficients are expected to be significant and should be included.

6.2. THE ELEMENTARY RATE OF ADSORPTION

The rate of adsorption is determined by the sticking coefficient. The sticking coefficient depends on the trapping probability of the molecule when it collides with the surface and the ratio of adsorption versus desorption probability of the trapped molecule. Trapping depends on the rate that translational energy can be dissipated by the surface or by molecules in the gas phase, and is related to thermal accommodation. For chemisorption, thermal accommodation is usually complete. Generally the rate of adsorption decreases with coverage, but exceptions exist.

The first step in a catalytic reaction is the adsorption of a molecule from the gas phase. When a molecule collides with the surface it can either adsorb or bounce back to the gas phase. We distinguish between two cases, trapping and sticking. If the attraction between the molecule and the surface is due to van der Waals forces, the molecule is physisorbed and the process is called trapping. If the attraction is stronger, a chemisorption bond may form. The sticking coefficient, s_0, is the probability that a molecule becomes chemisorbed at the surface.

Adsorption may proceed through a sequence of steps and the adsorbed species may be mobile or immobile. In general, the measured sticking coefficient corresponds to three successive processes (Figure 6.1): trapping of a molecule M into a precursor state M_p with a trapping constant k_t; desorption of M_p with a rate constant k_d; and transition from the precursor state to the chemisorbed state M_a with a rate constant k_a. As will be covered later, the precursor can be considered as a mobile state which is not associated with a particular site on the surface. These steps are summarized in the following set of reactions:

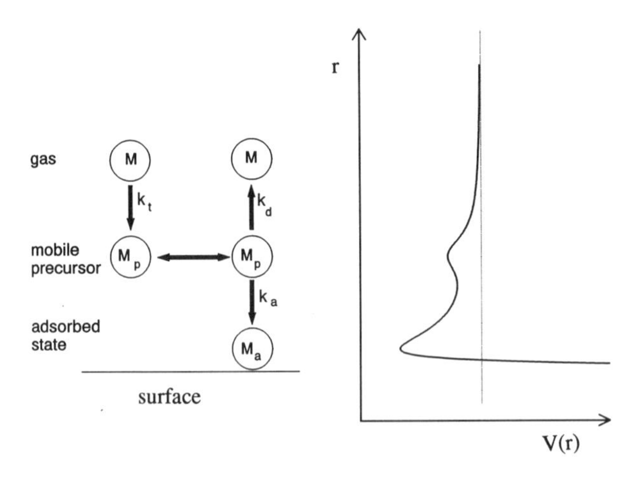

Figure 6.1. Adsorption via a precursor state and the corresponding representation in a potential energy diagram as a function of the distance r between adsorbate and surface.

$$M \underset{k_t}{\overset{k_d}{\leftrightarrows}} M_p$$

$$M_p \underset{k_a}{\rightarrow} M_a \tag{6.18}$$

The rate of adsorption follows from the kinetic equations

$$\frac{d[M_p]}{dt} = k_t[M] - k_d[M_p] - k_a[M_p](1-\theta)$$

$$r_a = k_a[M_p](1-\theta) \tag{6.19}$$

According to the steady-state assumption discussed in Chapter 2, $d[M_p]/dt = 0$, and thus the rate of adsorption is

$$r_a = \frac{k_a k_t}{k_a + k_d}[M](1-\theta) \tag{6.20}$$

The trapping rate constant per molecule is defined as

$$k_t = \alpha r_c = \frac{1}{4}\alpha\bar{u} \tag{6.21}$$

where r_c is the collision rate per unit surface area and u the average velocity of the gas phase molecules. Expression (6.21) follows if using (4.110) or (4.114). The trapping coefficient α is related to (but not the same as) the thermal accommodation coefficient γ discussed in Chapter 5. The former pertains to trapping into a precursor state. Remember that thermal accommodation is controlled by inelastic collisions and hence increases when a surface becomes covered with adsorbate molecules which are low in mass compared to the surface atoms. So the rate of adsorption of a molecule may increase with coverage when trapping by loss of energy is rate controlling.

With the definition given above, the following expression is found for the sticking coefficient on an empty surface ($\theta = 0$)

$$s_0 = \frac{r_a}{\frac{\bar{u}}{4}[M]} = \frac{k_a k_t}{k_a + k_d}\frac{4}{\bar{u}} \tag{6.22}$$

The relation between trapping coefficient and sticking coefficient becomes

$$s_0 = \frac{\alpha}{1 + k_d/k_a} \tag{6.23}$$

TABLE 6.2. Experimental Values of Sticking Probabilities of Gases on Single Crystal Surfaces[a]

Surface	Gas	T (K)	s_0
W(111)	N_2	300	0.004–0.08
W(100)	N_2	300	0.25–0.6
W(110)	N_2	300	$< 10^{-2}$
W(210)	N_2	300	0.28
W(310)	N_2	300	0.25–0.72
W(320)	N_2	300	0.73
W(110)	H_2	80	$< 10^{-4}$
W(110)	H_2	300	0.07
W(100)	H_2	300	0.18
W(111)	H_2	425	0.24
Re(0001)	N_2	300	< 0.002
Fe(110)	N_2	300	10^{-7}
Fe(100)	N_2	300	10^{-7}
Pt(111)	O_2	550	0.02
Pt(100)	O_2	300	0.1
Pt(110)	O_2	300	0.4
Pt(111)	H_2	150	0.016
Pt(111)	H_2	125	0.1
Pt(110)	H_2	125	0.33

[a] From Somorjai (1981).

The sticking coefficient of simple gases on bare transition metal surfaces varies in general between 10^{-3} and 1 (Table 6.2). The sticking coefficient depends on the trapping coefficient, which, in turn, is related to the process of thermal accommodation.

Chapter 4 defined the sticking coefficient in the transition-state theory in (4.183) as

$$s_{\mathrm{TST}} = \frac{k^T_{\mathrm{diss}}}{\bar{u}/4} \tag{6.24}$$

The relation between s_{TST} and s_0 becomes

$$s_0 = \alpha \cdot s_{\mathrm{TST}} \tag{6.25}$$

and can be written as

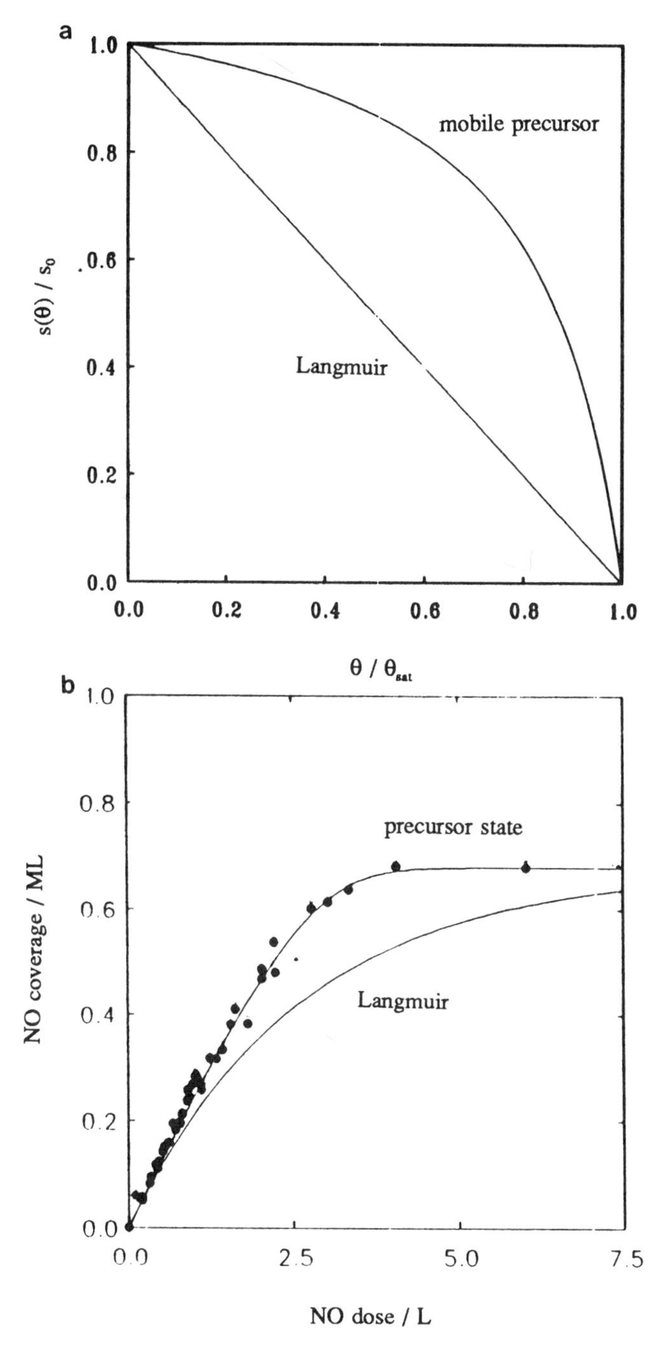

Figure 6.2. (a) Sticking coefficient as a function of coverage in the case of direct adsorption and a mobile precursor; (b) adsorption isotherm of NO on the (111) surface of rhodium showing the effect of the precursor mechanism (from Borg *et al.*, 1994).

$$s_0 = P\, e^{-E/kT} \tag{6.26}$$

When the pre-exponent of the adsorption rate is small due to loss of rotational degrees of freedom, or when the rate of trapping in the precursor state is low, s_0 will be small. Values of α close to 1 may be expected for molecules that have a large adsorption energy (see Chapter 5). Such molecules stay long enough at the surface for energy exchange to occur, so that accommodation can be achieved. This is the case for CO adsorption on metal surfaces. CO has a heat of adsorption in the order of 125 kJ/mol, which is at least an order of magnitude larger than RT. Lower sticking coefficients are expected for molecules that adsorb weakly, due to the low trapping coefficient. This is true for molecules such as CH_4 or noble gases. These molecules, which possess closed electron shells, interact only weakly with the surface.

In general the rate of adsorption is expected to decrease when the surface fills with adsorbates, because less sites become available for adsorption

$$r_a(\theta) = r_a(0)\cdot(1-\theta) - k_{des}\theta \tag{6.27}$$

This is the observed behavior when molecules adsorb directly. However, frequently the rate of adsorption decreases less steeply than predicted by (6.27), which is caused by the trapping of molecules in a mobile precursor state (Figure 6.2). The

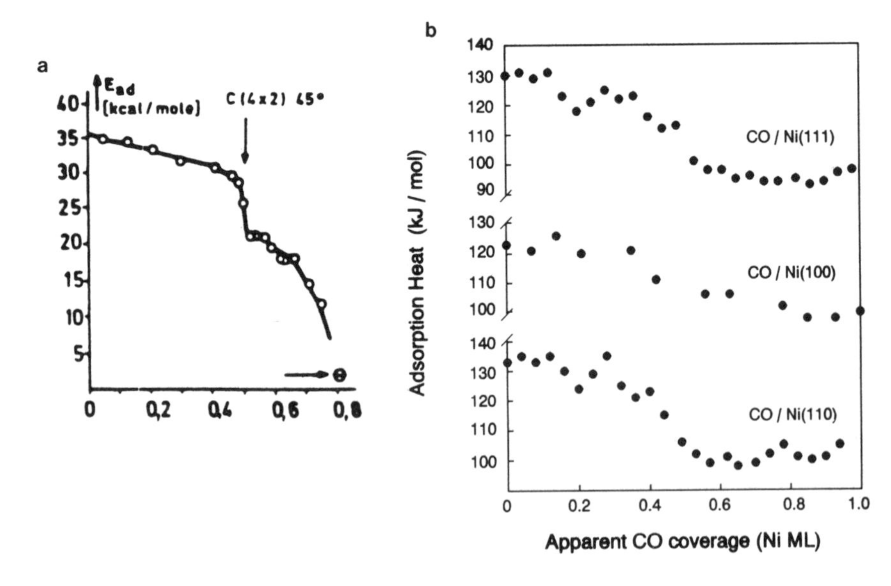

Figure 6.3. The heat of adsorption of CO on palladium and nickel surfaces depends on coverage (CO/Pd data from Conrad *et al.*, 1974, CO/Ni from Stuckless *et al.*, 1993).

adsorbing molecule may be accommodated on occupied sites and diffuses over the surface until it finds a free site where it can form a chemisorption bond. Sometimes the sticking coefficient increases initially with coverage when the coadsorbed species actually enhance the trapping coefficient by providing either a deeper attractive potential energy well or a larger collision probability due to the more corrugated surface.

Often, the heat of adsorption shows a sharp decrease when nearly half of the surface is covered with adsorbates, caused by lateral repulsive interactions between the adsorbates (Figure 6.3). It results in activated adsorption with a low sticking coefficient at high coverage. Low sticking coefficients often prevail under the typical conditions under which catalytic reactions run.

6.3. DISSOCIATIVE ADSORPTION

6.3.1. Introduction

> *At least one of the reactants in a catalytic reaction has to dissociate. Reduction of the energy barrier for dissociation is an important function of the catalytic surface. Closed shell molecules, such as H_2 and CH_4, interact before dissociation only weakly with the surface.*

A catalytic reaction is initiated when adsorbed molecular fragments are produced on the surface as a result of dissociation. Figure 6.4 shows two different representations for the change in potential energy which accompanies the dissociation of molecules. In Figure 6.4a the potential energy is plotted as a function of the two distances between the two atoms of the molecule and between the atom and the surface. The curves are lines of equal potential energy. This figure is a simplification since only the relative coordinate of the molecular center of mass

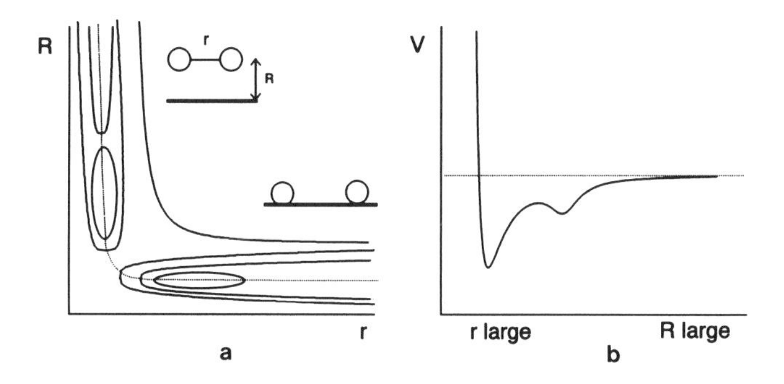

Figure 6.4. Potential energy diagrams for dissociative adsorption through a molecular precursor.

and the surface, R, and the separation between the two atoms, r, are considered and not the detailed geometry of the molecule with respect to the surface. In general the molecule can approach the surface in any orientation. Clearly, for dissociation the parallel position is required. Computational studies (see Section 6.3.5.1) have shown that molecules rotating in a plane parallel to the surface have a preference for dissociation.

Figure 6.4b shows the potential energy along the most probable reaction path of Figure 6.4a. The potential has a shallow minimum corresponding to the physisorbed state, at a relatively large distance R from the surface, where the intramolecular distance r is close to the equilibrium value of the molecule in the gas phase. The forces responsible for physical adsorption are the weak van der Waals forces. These dispersive forces are proportional to the polarizability of a molecule and result from electromagnetic interactions between the adsorbing molecule and the surface.

6.3.2. Molecular Orbital Picture of Chemisorption

The activation energy is the result of two different types of interactions. The bond length within the dissociating molecule increases upon adsorption. This process costs energy. The orbitals of adsorbate and surface need to have finite overlap in order for a covalent surface bond to form. The covalent interaction with the metal surface lowers the energy cost of dissociation.

In order to analyze the interaction between a molecule and the catalytic surface that leads to molecular or dissociative chemisorption, it is useful to consider the metal-adsorbate system as a "surface molecule." This complex has molecular orbitals, composed of metal orbitals and orbitals of the adsorbing molecule.

The adsorption of a single atom is considered first. Figure 6.5 gives schematic orbital diagrams. The overlap between the electron densities of the metal and the adsorbing atom gives rise to the formation of a new pair of broad orbitals which are occupied by electrons from the atom and from the electron bands of the metal. An attractive interaction occurs if the bonding orbital is occupied, while occupation of the antibonding orbital results in a weak bond. Several situations can arise:

1. The antibonding chemisorption orbital falls entirely above the Fermi level, it remains empty and a strong chemisorption bond results (Figure 6.5a).
2. If the interaction between the atomic orbital and the metal orbitals is weak, the energy splitting between the bonding and the antibonding-orbital fragments of the chemisorption bond is small. The antibonding orbital falls below the Fermi level and becomes occupied. This does not lead to bonding but to repulsion and the atom leaves the surface (Figure 6.5c).

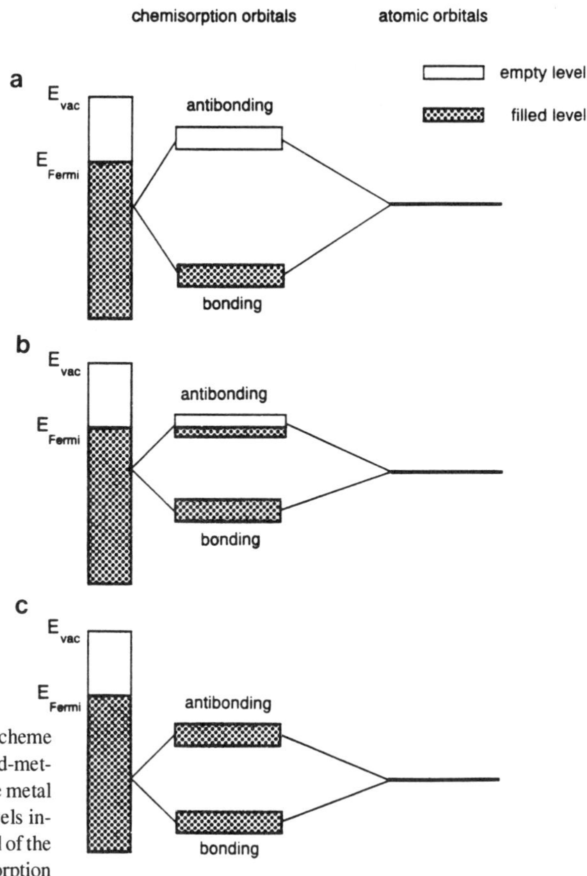

Figure 6.5 Simplified orbital scheme for chemisorption of atoms on d-metals; left) the electron band of the metal with the Fermi and vacuum levels indicated; right) the atomic orbital of the free atom; and middle) the adsorption orbitals.

3. Intermediate situations in which the antibonding chemisorption orbital broadens across the Fermi level can also arise (Figure 6.5b). In such cases the antibonding orbitals are only partially filled and the atom A will be chemisorbed, however with a weaker chemisorption bond than in Figure 6.5a.

To understand the conditions under which a chemisorbed diatomic molecule such as H_2, N_2, or CO dissociates, two orbitals of the molecule need to be taken into account, the highest occupied and the lowest unoccupied molecular orbital (the HOMO and LUMO within the frontier orbital concept). Consider this simple case: the molecule A_2 with doubly occupied bonding level σ and unoccupied antibonding level σ^*. As in H_2 the interaction of each of these levels of the molecule with the

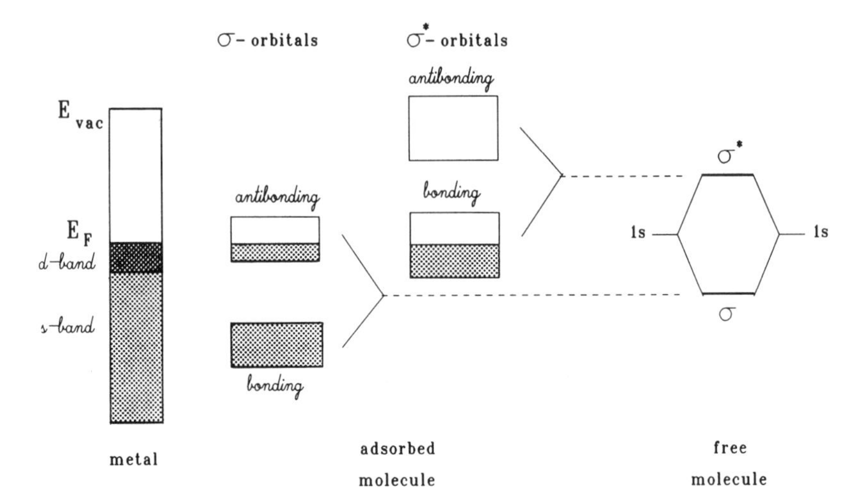

Figure 6.6. Orbital scheme for chemisorption of a diatomic molecule on a d-metal. Orbitals are formed from the bonding and antibonding levels of the molecule. Partial filling of the latter weakens the intramolecular bond (from Niemantsverdriet, 1993).

d- and s-levels of the metal must now be considered. These are the necessary steps (see Figure 6.6):

1. Construct new molecular orbitals from the HOMO, in this case the bonding orbital σ of A_2, and a level in the surface that has appropriate orientation and symmetry.
2. Do the same for the LUMO, here the antibonding σ^* of A_2, and another level of appropriate orientation and symmetry.
3. See where the levels are with respect to the Fermi level of the metal, and find out which of the orbitals are filled and to what extent.

What are the important things to look at? First, the interaction under (a) between the occupied molecular σ-orbital and an occupied surface orbital gives in principle a repulsive interaction, because both the bonding and the antibonding chemisorption orbital will be occupied. However, if the antibonding orbital falls above the Fermi level, the repulsion is partially or entirely relieved (as in the interaction of the 5σ-orbital of CO on rhodium metal). Second, interaction (b) gives a bonding orbital which can be either above or below the Fermi level. Since the participating LUMO orbital of the adsorbing molecule is antibonding with respect to the interaction between the molecule atoms, occupation of the corresponding orbital fragments leads to dissociation of the molecule. If it is partially occupied, then it contributes less to bonding between A_2 and the surface, while at the same time the

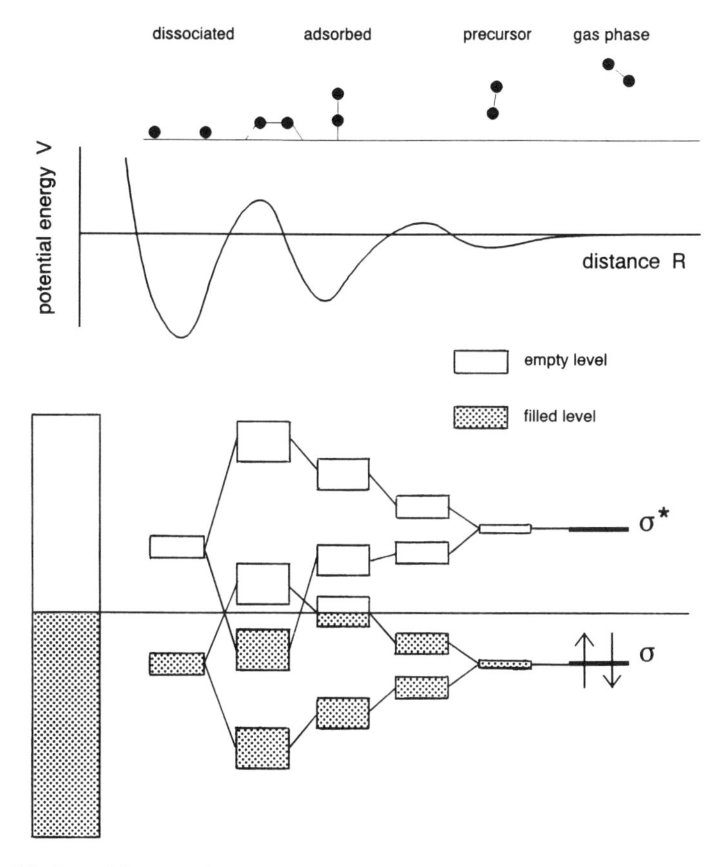

Figure 6.7. Potential energy diagram for the approach of a diatomic molecule from the gas phase to the surface along with the corresponding orbital diagrams.

intramolecular A–A bond of the chemisorbed molecule is weakened (as is the case with the $2\pi^*$-orbital of CO on most Group VIII metals).

It is instructive to make the connection between potential energy diagrams for chemisorption and the molecular orbital picture just discussed (Figure 6.7). At the right of the potential energy diagram the molecule is in the gas phase, where its levels are unperturbed. Closer to the surface, the molecule is trapped in the physisorbed state. Its wave functions have very little overlap with the electrons of the metal and the levels only slightly broadened. This weakly bonding state is due to van der Waals interactions. At smaller distances, the wave functions of the molecular levels and electron levels from the metal surface start to overlap and bonding and antibonding molecular orbital fragments between surface and adsorbate start to form. However, as the orbital overlap is still small, both the antibonding and the bonding interaction levels fall below the Fermi level and are occupied. This creates repulsive interaction between the surface and the adsorbing atom. This causes part of the increase in potential energy which generates an activation barrier.

As the wave function overlap increases, the antibonding orbitals go up in energy, shift above the Fermi level, and become empty. Repulsion is relieved. At the same time the bonding orbital has shifted down in energy, and a strong chemisorption bond has developed. In the situation of Figure 6.7, the molecule stays intact because the chemisorption orbital formed from the σ^* level remains above the Fermi level. If the molecule comes closer to the surface, the stronger interaction broadens the σ^*-derived levels and pushes the corresponding bonding-orbital levels down, such that they become partially occupied by electrons which come originally from the metal. Because this process leads to a population of the antibonding σ^*-molecular orbital this process is often called "back donation." The partial filling of levels, which are bonding for the metal-molecule interaction but antibonding for the adsorbed molecule, creates a favorable situation with an activation barrier for dissociation much lower than that for dissociation of the molecule in the gas phase.

Potential energy representations as in Figure 6.7 are conceptually useful but also have limitations. Not only the distance between the center of gravity of the molecule and the surface, R, but also the distance between the constituent atoms of the molecule, r, is important, and should be considered. As sketched in Figure 6.4, the intramolecular bond stretches when the molecule approaches the metal surface. This implies that the energy difference between the bonding and antibonding molecular orbitals decreases. Back donation delivers an electron to the molecule, antibonding (i.e. bond weakening) orbitals become occupied and the molecule expands further. This is the origin of the low activation energy for dissociation in Figure 6.4. The closer the molecule can come to the surface, the higher electron back donation becomes and the more the intramolecular bond stretches.

6.3.3. Chemistry of Surface Dissociation

Whereas the bond strength of adsorbed atoms varies strongly with the metal substrate, the adsorption bond strength of molecules varies much less. Within the same row of the periodic system, transition metals with less d-valence electrons are more reactive. Open surfaces, with atoms of decreased coordination, are more reactive than the more dense surfaces.

The analysis of the surface–chemical bond in the previous subsection was given in terms of the electron occupation of the bonding and antibonding states of the molecular orbitals between the adsorbate and the surface. This concept will be used to discuss trends in the bond strength of adatoms on a transition metal as a function of its position in the periodic system.

When comparing the electronic structure of transition metals in a row of the periodic system, the d-valence-electron occupation increases with increasing atomic number. The d-electron bands are completely filled in the Group I-B elements, Cu, Ag, and Au. Moving to the left along the row in the periodic system, the heat of adsorption of atoms goes up, provided one compares substrate surfaces of the same geometry. Two factors are responsible. First, the depletion of the d-band causes a decrease in the occupation of the antibonding electron density between adsorbate and surface. This results in an increasing dominance of the bonding contribution to the chemical bond. Second, the ionization potential of the elements decreases moving towards the left along a row of the periodic system, resulting in an enhanced ionicity of the surface chemical bond.

Another trend is that the bond strength between adsorbates and metals decreases when moving down in the same column of the periodic system. This is mainly because the metallic bond between heavy metal atoms is stronger than between light metal atoms. As adsorbate electrons compete for covalent bonding with the electrons of the metal, the adsorption bond tends to be weaker for chemisorption on heavier metals. It is useful to remember these two general trends when comparing the catalytic activity of different metals in the next section.

The adsorption bond strengths of molecules vary in general much less with the substrate than those of adsorbed atoms. This is particularly true in the case of molecules for which the LUMO differ little in energy with the highest occupied levels in the electron band of the metal.

As already mentioned, the way electrons of the metal interact with the HOMO and LUMO of a molecule is qualitatively very different. Bonding levels formed from the metal band and the HOMO are filled, as are a substantial fraction of the antibonding component. In contrast, the antibonding levels formed from the metal band and the LUMO of the adsorbed molecule remain empty, and the bonding levels are only partially occupied. Hence, when the total electron concentration between adsorbate and metal surface decreases, as is the case when the d-valence-electron occupancy goes down, the interaction with the LUMO of the adsorbate decreases, whereas the interaction with the HOMO increases. The total bond strength varies little due to the two counteracting effects. Usually the change in interaction with the HOMO will tip the balance.

In contrast to molecular adsorption, the interaction energy of atoms as C, O, or N with a metal surface is a strong function of coordination number. Adsorbed atoms almost always prefer bonding in sites of threefold or fourfold coordination. Molecules can also adsorb on top or bridge sites. As will be seen, this has major consequences for the dissociation paths of diatomic molecules on metal surfaces.

Whether or not an adsorbed molecule dissociates is determined by the thermodynamics of the adsorption states before and after reaction and the energy barrier between them. The major difference between dissociation in the gas phase and the adsorbed state is that dissociation fragments are stabilized by adsorption to the

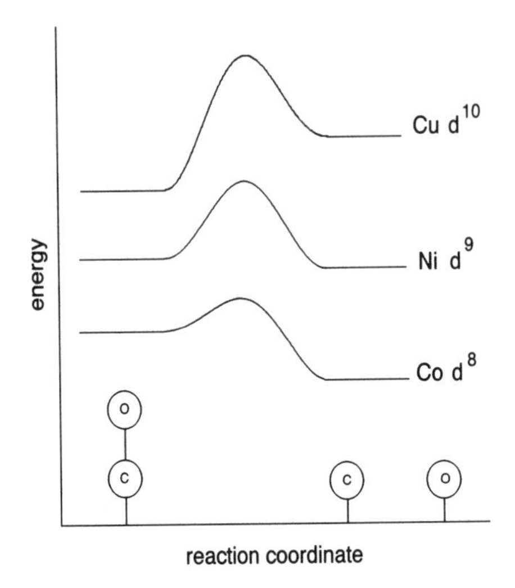

Figure 6.8. Schematic potential energy diagram for the dissociation of CO on metals with a different number of d-electrons.

catalysts surface. The interaction energies of atoms change more strongly with the substrate metal than do those of molecules. Figure 6.8 shows schematic energy diagrams for the dissociation of CO on 3d-metals. The dissociation of CO is endothermic on copper, slightly exothermic on nickel, and highly exothermic on cobalt and iron. Dissociation becomes progressively more unfavorable for 4d- and 5d-metals if staying within the same column of the periodic system. Note that the activation energy for CO dissociation decreases along with the increase in heat of reaction, in agreement with the Polanyi–Brønsted relation discussed in Section 5.6, Expression (5.53) implies that the change in activation energy, ΔE_{act}, is roughly proportional to half the change in heat of reaction, ΔH_{react},

$$\Delta E_{act} \approx \frac{1}{2}\Delta H_{react} \tag{6.28}$$

The adsorption geometry of the atoms produced when the molecule dissociates determines the size of the ensemble of surface atoms needed for the dissociation. As sketched in Figures 6.9 and 6.10, carbon monoxide requires at least five atoms

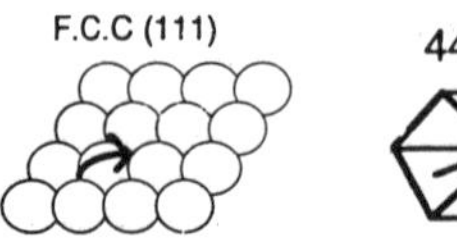

Figure 6.9. Most probable dissociation path of CO on a rhodium (111) surface (from de Koster and van Santen, 1990).

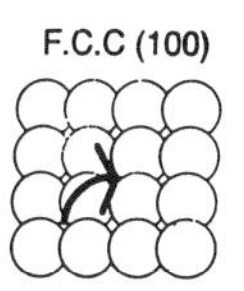

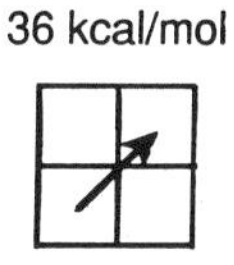

Figure 6.10. Most probable dissociation path of CO on a rhodium (100) surface (from de Koster and van Santen, 1990).

on the (111) surface of an fcc transition metal in order to dissociate, and on the fcc (100) surface seven to eight atoms. Dissociation proceeds by bending the CO molecule across an atom. In this way, the antibonding CO orbitals have a strong interaction with the d-orbitals of the metal, and the C–O bond is efficiently weakened.

The activation energy for CO dissociation on rhodium is lower for the (100) than for the (111) surface. The (100) surface is the more open one of the two. Each rhodium atom misses four nearest neighbors in the first coordination shell as compared to a bulk atom, whereas an atom in the (111) surface misses three nearest neighbors. Hence, carbon and oxygen atoms bond more strongly with the (100) than with (111) surface and, similar to the situation in Figure 6.8, both thermodynamics and kinetics are more favorable for CO dissociation on the (100) surface. This is a manifestation of the bond order conservation principle: *The more the valence electrons of a metal atom at the surface become distributed in bonds with neighboring atoms, the weaker the individual bond becomes.*

The increased reactivity of small particles has the same basis. The atoms at the surface of small particles have a low coordination number and therefore a high reactivity. Dissociation becomes more favorable, provided the particles are large enough for their exposed surfaces to expose the ensembles of metal atoms needed to dissociate a molecule.

The bond order conservation principle is conceptually useful but only approximately correct. Changes in ionization potential and electron affinity are also important and have to be included in more rigorous considerations. Nonetheless, to a first approximation changes in reactivity are primarily related to the coordination and geometry in the first coordination shell of surface atoms.

6.3.4. Chemical Precursor-Assisted Dissociation

Short-lived chemical precursors may lower the activation energy for dissociative adsorption.

In some cases it has been observed that the dissociation of adsorbed molecules is facilitated by the presence of a second gas. Remarkably, the second gas does not necessarily have to be present in a chemisorbed state. Au and Roberts (1986) observed an enhancement of the rate of reaction of NH_3 on Zn or Cu surfaces by the presence of O_2 in the gas phase. They suggested that ammonia decomposition

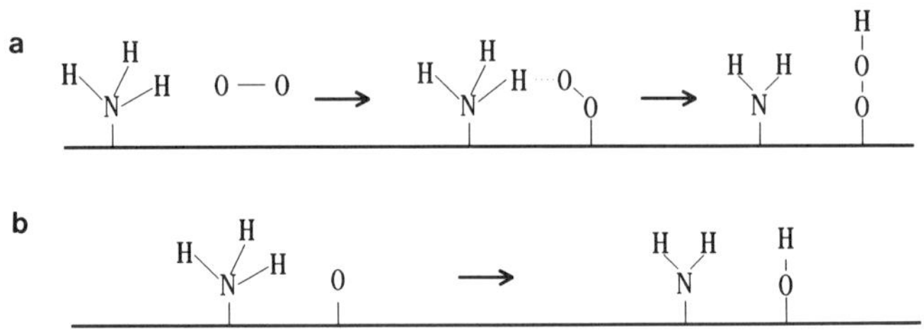

a

b

Figure 6.11. Reaction between adsorbed ammonia and atomic (bottom) or molecular oxygen (top). The latter is to be considered a chemical precursor.

may be enhanced by the formation of an intermediate surface complex as sketched in Figure 6.11a. Theoretical calculations confirm the feasibility of such a mechanism as we explain in the following.

On metal surfaces, molecularly adsorbed oxygen is only stable at low temperatures (i.e., below 150 K); it dissociates at higher temperatures. Under the conditions of Roberts' experiment, the O_2 molecule will rapidly dissociate when adsorbed on the metal surface, after which a reaction according to Figure 6.11b would be expected. However, the strong N–H bond in NH_3 and the strong Cu–O bond of atomic oxygen on copper make this reaction endothermic by 50 kJ/mole, and thus unlikely, see Neurock *et al.*, 1994.

Calculations on the reaction between adsorbed ammonia and the O_2 molecule in a physisorbed precursor state as sketched in Figure 6.11a indicate that this mechanism is thermodynamically feasible. The O–O bond is significantly weaker than the Cu–O bond, and consequently the reaction of Figure 6.11b is exothermic with an overall heat of reaction of 80 kJ/mol. The OOH_{ads} peroxide decomposes in OH_{ads} and O_{ads}, and the latter reacts rapidly with $NH_{2,ads}$ and NH_{ads}. Such reactions are thermodynamically much more favorable than the one between $NH_{3,ads}$ and O_{ads} because of the much weaker N–H bond in these species.

The crux of the mechanism is that the reaction of adsorbed ammonia with molecular oxygen in a precursor state is faster than the dissociation of the oxygen. Since the molecular oxygen is a transient species under the conditions of this reaction, it is to be considered a chemical precursor. Chemical precursors are very difficult to identify spectroscopically, but have incidentally been invoked in catalytic reaction mechanisms, such as the total oxidation of methane. The rate-limiting step of this reaction is the breaking of the first C–H bond of CH_4; this step may be facilitated by the interaction with molecular oxygen present in a precursor state.

6.3.5. Dynamics of Dissociation; Quantum-Mechanical Tunneling

6.3.5.1. The Dissociation of Hydrogen by Copper

Details of dissociation dynamics can be probed using scattering experiments. Vibrational excitation of the molecular bond lowers the translational energy required for dissociation upon adsorption. Also, the relative orientation of the molecule with respect to the surface is important for dissociation.

Transition-state theory assumes rapid energy exchange between reacting molecules and their environment. The details of the motion of the reaction coordinate are only important in the stage where the activation energy barrier is crossed; they determine the probability that the system actually goes over the barrier. In the conventional transition-state theory, discussed in Chapter 4, this probability has been taken equal to one.

Molecular beam experiments, in which molecules of well-defined translational and vibrational energy can be scattered by surfaces, probe details of the reaction path across the potential energy surface. Here we discuss the dynamics of hydrogen dissociation based on theoretical calculations which simulate the collisions of molecules with a surface as occurring in molecular beam experiments. A potential energy diagram for the H_2 dissociation is shown in Figure 6.12.

The two important coordinates are the distance of the H_2 center of mass with respect to the metal surface and the distance between the hydrogen atoms. As Figure 6.12 indicates, the H–H bond stretches when the molecule comes close to the surface; the barrier to dissociation occurs when the molecule is close to the surface and has stretched considerably. For this reason it is called a "late" barrier, whereas early barriers are found in the entrance channel of the process. The fact that H_2 dissociation is associated with a late barrier has consequences for the way the sticking coefficient for dissociative adsorption depends on energy. This is illustrated by the results shown in Figure 6.13.

Sticking coefficients for dissociative adsorption of hydrogen may be low, especially on copper where the process is activated. Measurements by Rasmussen *et al.* (1993) indicate values for s_0 as low as 10^{-14}–10^{-13} for H_2 on Cu (100). Only sufficiently energetic molecules will be able to overcome the barrier for dissociation. Figure 6.12 illustrates that not only the translational, but also the internal energy of the molecule increases the dissociation probability of hydrogen. Note that a molecule in the vibrationally excited state needs less translational energy than does a molecule in the ground state. This agrees with the potential energy diagram of Figure 6.12, which indicates that the H–H bond has to stretch before it crosses the activation barrier for dissociation.

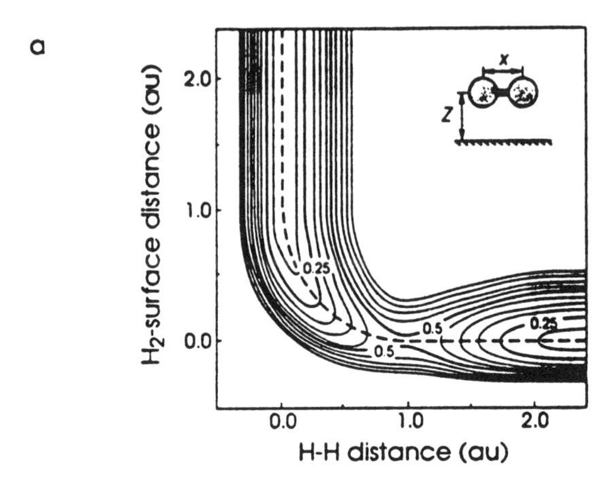

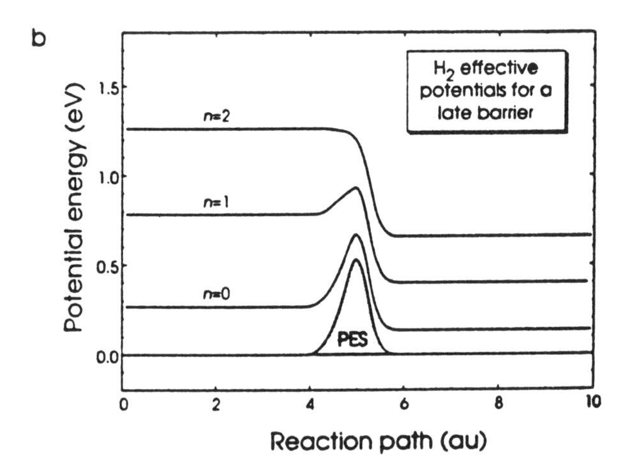

Figure 6.12. (a) Potential energy surface for H_2 approaching a Cu(111) surface as shown in the inset; (b) potential energy along the reaction path in (a) for different vibrational states of the molecule (from Holloway and Darling, 1992).

Increased rotational energy helps to dissociate the molecule, as Figure 6.13 shows. Calculations indicate that rotational motion in a plane perpendicular to the surface (the "cartwheel" mode) is restricted, because it affects the interaction between the molecule and the surface. Rotation parallel to the surface (the "helicopter mode") is virtually free, however, and allows for stretching of the H–H bond as required for dissociation.

Calculations as the ones underlying the results of this section are done by solving the time-dependent Schrödinger equation. Quantum-mechanical tunneling phenomena, to be discussed in the next section, are responsible for dissociation at

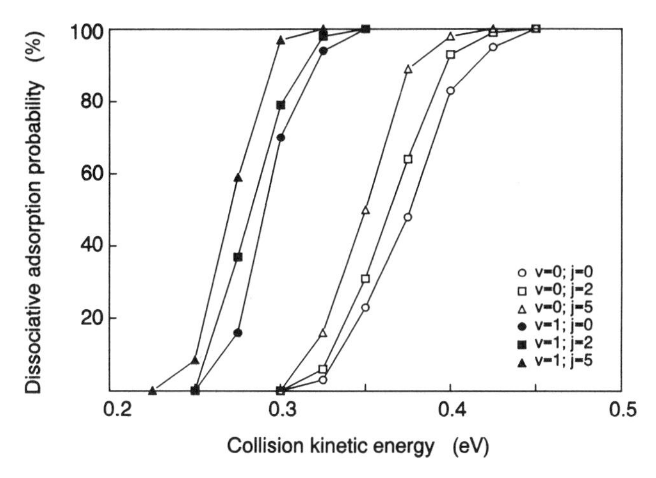

Figure 6.13. Sticking coefficient of dissociative hydrogen adsorption on Fe(100) as a function of translational energy for different vibrational and rotational states of the molecule (from Tantardini, 1992).

energies that would be forbidden according to classical mechanics. Replacement of hydrogen by deuterium changes the effective barrier height by the altered zero vibrational energy, $\frac{1}{2}h\nu$, because the vibrational frequency depends on the mass of the vibrating atoms. The tunneling probability also depends on the mass, (6.34). The two effects result in a sticking coefficient for D_2 that is lower than for H_2.

The late barrier with the elongated H–H bond should, according to the formalism of Chapter 4, be qualified as a loose transition state, in which the molecule is even allowed to rotate in a plane parallel to the surface. Interestingly, we will also find a loose transition state for the dissociation of methane, to be discussed in Section 6.4.2.

As the theoretical treatment of reactions involving hydrogen needs to take the quantum mechanical nature of the light hydrogen atom into account, the implications of quantum-mechanical tunneling processes through the activation barrier will be covered first.

6.3.5.2. The Tunneling Correction to the Rate Constant in the Transition-State Theory

Tunneling through the energy barrier increases the rate of reactions involving hydrogen. Isotope effects are caused by a change in zero-point vibrational frequencies and by altered tunneling probabilities.

The small mass of the hydrogen atom causes its motion to be non-classical. It is also the reason vibrational frequencies of bonds with hydrogen are high. The quantum mechanical nature of the proton implies it has both particle and wave character. It is the latter property that is responsible for the tunneling of reactions through instead of over the activation barrier. The contribution of tunneling will be estimated to the rate constant within the approximation that the motion of the hydrogen atom is one dimensional.

The Schrödinger equation to be solved in one dimension is

$$\left(-\frac{\hbar^2}{2m}\frac{d^2}{dx^2}+V(x)\right)\psi(x)=\varepsilon\,\psi(x) \tag{6.29}$$

Taking for V the rectangular barrier of Figure 6.14, the Schrödinger equation reduces to

$$-\frac{\hbar^2}{2m}\frac{d^2}{dx^2}\psi(x)=\varepsilon\,\psi(x) \qquad x<0;\, x>d$$

$$\left(-\frac{\hbar^2}{2m}\frac{d^2}{dx^2}+U\right)\psi(x)=\varepsilon\,\psi(x) \qquad 0<x<d \tag{6.30}$$

Thus there are three regions, where the solutions have the general forms

$$\psi_1(x)=Ae^{iax}+Be^{-iax}$$

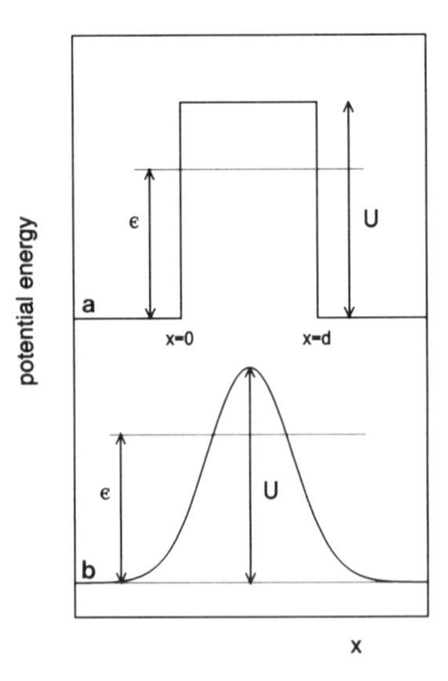

Figure 6.14. Potential energy barriers for the transition state with a total energy ε of an approaching particle; the lower curve represents an Eckart barrier.

$$\psi_2(x) = Ce^{\beta x} + De^{-\beta x}$$

$$\psi_3(x) = Ee^{iax} \tag{6.31}$$

The first wave function, ψ_1, represents the sum of the waves coming in from the left and the part that is reflected by the barrier. The function ψ_3 is the wave transmitted through the barrier and ψ_2 is the tunneling wave. From (6.30) there is

$$\alpha = \frac{2\pi}{h}\sqrt{2m\varepsilon}\,; \quad \beta = \frac{2\pi}{h}\sqrt{2m(U-\varepsilon)} \tag{6.32}$$

The coefficient beta is only real if $\varepsilon < V$. Using the continuity conditions for ψ and its derivative at the beginning and end of the barrier ($x = 0$ and $x = d$), the transmission probability for tunneling, $F(\varepsilon)$ is

$$F(\varepsilon) = \frac{|\psi_3|^2}{|\psi_1|^2} = \left|\frac{E}{A}\right|^2 = \frac{1}{\cosh^2\beta d + \dfrac{(\alpha^2-\beta^2)^2}{4\alpha^2\beta^2}\sinh^2\beta d} \tag{6.33}$$

The tunneling probability approaches 1 for high energies. When the energy ε is low compared to the height of the barrier, the transmission probability becomes

$$F(\varepsilon) \approx e^{-(4\pi\sqrt{2m(U-\varepsilon)})/h}\,; \quad \varepsilon \ll U \tag{6.34}$$

Thus, for small energies of the tunneling particle, the tunneling probability decreases exponentially with the width of the barrier, but less steeply with its height. Tunneling also becomes less probable with increasing mass of the particle, as expected.

In Chapter 4 we deduced (4.146) and (4.151) for the rate constant in the transition-state theory. Quantum mechanical effects were ignored. To take tunneling into account, the rate constant is written as

$$\begin{aligned} k_T^{\Gamma} &= k_T \cdot \Gamma \\ &= k_T \cdot \frac{e^{U/kT}}{kT}\int_0^{\infty} F(\varepsilon)e^{-\varepsilon/kT}d\varepsilon \end{aligned} \tag{6.35}$$

where U is the barrier height (Figure 6.14a), but corrected for the zero-point vibrational energies of reactants in the ground state and the transition state.

Instead of using a rectangular barrier, the tunneling probability can also be calculated for more realistic potentials, such as the Eckart potential of Figure 6.14a. Trends in $K(\varepsilon)$ remain qualitatively the same: The tunneling probability increases with decreasing barrier width and with increasing energy of the incoming particle.

Note that classically $K(\varepsilon)$ is a step function which is zero for energies smaller than the barrier height and unity for higher energies (Figure 6.14b). Finally in reactions where quantum mechanical tunneling is likely, the transmission coefficient Γ is typically between 1 and 2, although higher values may arise if energy barriers become unusually narrow.

6.4. TRANSITION STATES OF SURFACE REACTIONS

Transition states for reactions on surfaces can be predicted by using ab-initio *quantum-chemical calculations in which the surface is represented by a small cluster.*

In Table 4.9, experimental data of rates of desorption and dissociation were compared, and it was concluded that the preexponential factors for dissociation are often a few orders of magnitude smaller than those for desorption. Two kinds of transition states were defined: loose and tight. A loose transition state offers considerable freedom to the reacting species, whereas the motion of the reacting fragments is significantly constrained in a tight transition state. The latter is associated with a low preexponent of the rate constant.

Prediction of the barrier energies as well as the vibrational and rotational motions of a transition state is currently becoming feasible on the basis of first-principles, *ab-initio* quantum-chemical calculations. Such calculations require that the actual surface is reduced to a cluster of atoms interacting with the reacting molecule. The use of this approach in the activation of methane on two surfaces will be illustrated. The dissociation of methane on a single nickel atom will be discussed first and the exchange between a hydrogen atom of methane with an acidic proton on a cluster representing the active site of a zeolite.

6.4.1. The Transition State for Methane Dissociation on a Metal

The interaction of a molecule with a metal surface weakens intramolecular bonds and facilitates the dissociation by giving rise to a low activation energy for bond breaking. C–H bond activation in adsorbed hydrocarbons involves the filling of antibonding orbitals by electrons from the metal while the C–H bond stretches. Calculations predicts that the transition state for the formation as well as the dissociation of C–H bonds has rotational freedom and should be characterized as a loose transition state.

The chemisorption of methane on a metal surface requires that one of the C–H bonds of CH_4 is broken

$$CH_4 + 2 * \rightarrow CH_{3,ads} + H_{ads} \tag{6.36}$$

Similarly, as examined in Section 6.3.3 for the dissociation of CO on rhodium, the activation of the C–H bond proceeds through a transition state in which a metal atom is crossed by the dissociating bond. Calculations of the potential energy surfaces for this reaction indicate that the geometry of the transition state for dissociation on a single nickel atom or on a nickel atom embedded in a larger cluster is very similar. The height of the energy barrier for dissociation, however, differs significantly and reflects mainly the difference in reactivity between a single atom and an atom coordinated to other metal atoms.

The chemical features determining the potential-energy path for dissociation of the C–H bond are lowering of the activation energy by the interaction of orbitals on Ni with atomic orbitals on carbon and hydrogen. Population of antibonding orbital fragments while the C–H bond stretches, weakens this bond, and lowers the activation energy for bond breaking. These antibonding orbital fragments are empty in the free CH_4 molecule but become occupied as soon as a chemical bond develops with the atomic orbitals of the metal.

Figure 6.15 presents the geometry of the transition state for the dissociation of methane by a single nickel atom as it is predicted by density functional calculations. H_{act} is the atom that dissociates from the CH_4 molecule. The adsorption reaction can actually be written as the second-order process

$$CH_4 + Ni \xrightarrow{k_{diss}} H\text{–}Ni\text{–}CH_3 \tag{6.37}$$

which is accompanied by the reverse, first-order desorption process

$$H - Ni - CH_3 \xrightarrow{k_{des}} CH_4 + Ni \tag{6.38}$$

In Figure 6.15, the vibrational frequencies in the transition state have been indicated. These follow directly from the shape of the potential energy curves characterizing the bond. In case of the harmonic potential, $V(r) = \frac{1}{2} f \cdot (r - r_{eq})^2$, the vibrational frequency is given by

$$\nu = \frac{1}{2\pi}\left(\frac{f}{\mu}\right)^{1/2}; f = \frac{\partial^2 V}{\partial r^2}\bigg|_{r = r_{eq}} \tag{6.39}$$

in which f is the force constant of the bond and μ the reduced mass of the participating atoms. Expression (6.39) remains valid in the case of other potential energy functions.

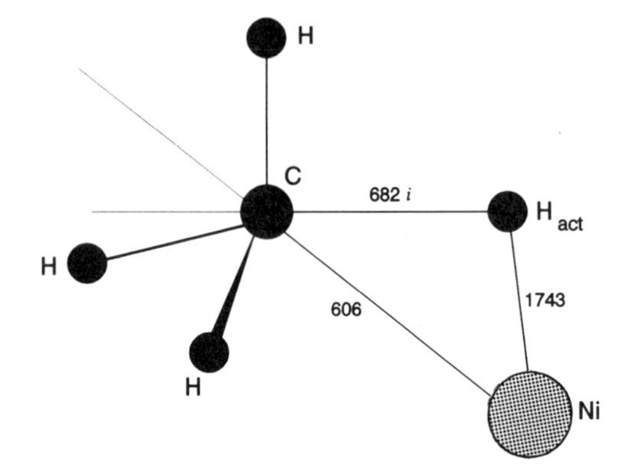

Figure 6.15. Transition state of CH_4 dissociation on a nickel atom. Characteristic frequencies for the stretch vibrations of the bonds have been indicated; the imaginary frequency represents the reaction coordinate (from Burghgraef *et al.*, 1993).

Note that one of the modes (the C–H_{act} stretch) is imaginary. This is the reaction coordinate. Because it is the solution of the equation of motion, computed for the geometry of a molecule at the saddle point of the energy barrier, it is the solution of Newton's equation with a negative force constant. Because the second derivative of the potential energy surface in the transition state with respect to the reaction coordinate is negative, the potential energy of the reaction barrier can be approximated by a negative parabola (Figure 6.16). According to (6.39), the frequency corresponding to the reaction coordinate becomes imaginary. Whereas real frequencies correspond to oscillations, imaginary frequencies correspond to motions that decay with time

$$A(\nu) \propto \mathrm{Re}\, e^{2\pi i \nu t} \propto 2\cos(2\pi\nu t) \qquad \text{if } \nu \text{ is real}$$

$$A(\nu) \propto \mathrm{Re}\, e^{2\pi i \nu t} \propto e^{-\lambda t} \qquad \text{if } \nu \text{ is imaginary; } 2\pi\nu = i\lambda \tag{6.40}$$

in which $A(\upsilon)$ is the amplitude of a normal mode.

The expressions for the rate constants of dissociation and desorption in the transition state are

$$k_{\mathrm{diss}} = \Gamma V \frac{kT}{h} \frac{pf^{\#}_{\mathrm{vib}}\, pf^{\#}_{\mathrm{rot}}}{pf_{\mathrm{trans}} pf_{\mathrm{vib}}\, pf_{\mathrm{rot}}} e^{-E_b/kT} \tag{6.41}$$

$$k_{\mathrm{des}} = \Gamma \frac{kT}{h} \frac{pf^{\#}_{\mathrm{vib}}\, pf^{\#}_{\mathrm{rot}}}{pf_{\mathrm{vib}}\, pf_{\mathrm{rot}}} e^{-E_b/kT} \tag{6.42}$$

where Γ is the transmission coefficient and E_b is the classical energy at which reaction occurs. The transmission coefficient is larger than 1, because the rate is

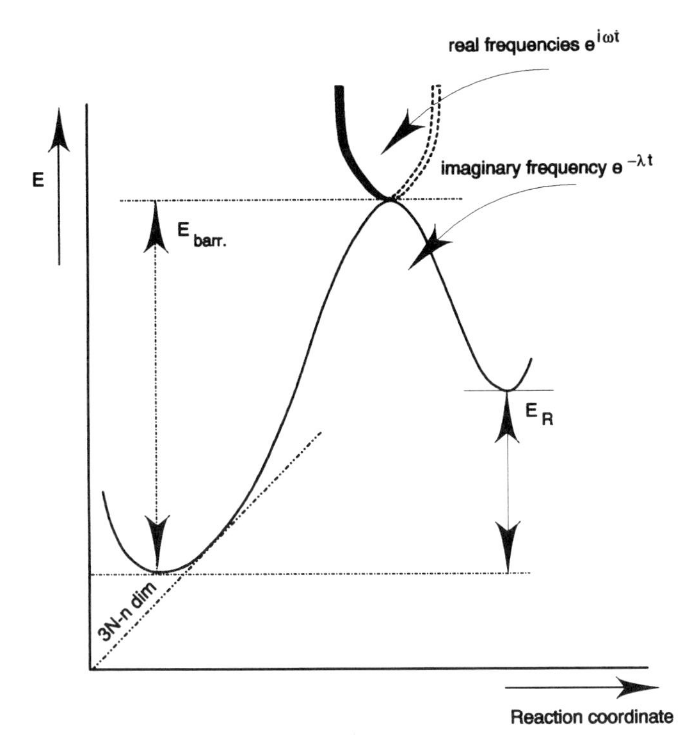

Figure 6.16. The reaction coordinate corresponds to the imaginary frequency associated with the *col* of the transition state.

enhanced by quantum-mechanical tunneling through the barrier, as discussed in Section 6.3.5.2. Note that the partition functions in the above expressions have been taken with respect to the bottoms of the potential wells. To compute the activation energy, the barrier energy has to be corrected for the zero-point energy contributions of transition and ground states, to give E_{crit}

$$E_{\text{crit}} = E_b - \frac{1}{2}\sum_i h\nu_i \tag{6.43}$$

Next the Arrhenius activation energy is calculated by applying (4.131).

Table 6.3 collects the computed values of E_b, E_{crit} as well as the activation energies and preexponential factors for the dissociation and desorption of CH_4 and CD_4 on the nickel atom. The activation energy has been computed from the rate constants using expression (4.131b). Table 6.4 gives the sticking coefficients, calculated using the "hard sphere" preexponential

TABLE 6.3. Characteristic Energies and Preexponential Factor for the Adsorption and Desorption of Methane on and from a Nickel Atom[a]

Molecule	Reaction	E_b (kJ/mol)	E_{crit} (kJ/mol)	E_{act} (kJ/mol)	ν_T[b]
CH_4	adsorption	40.7	34.8	36.2	$7.62 \cdot 10^{07}$
CD_4	adsorption	40.7	37.8	39.4	$5.56 \cdot 10^{07}$
CH_4	desorption	74.7	70.7	73.0	$1.16 \cdot 10^{13}$
CD_4	desorption	74.7	71.9	73.8	$9.43 \cdot 10^{12}$

[a]From Burghgraef *et al.* (1993).
[b]Unit for adsorption: $m^3/mol \cdot s$; unit for desorption: s^{-1}

$$\nu_{HS} = \pi r^2 \left(\frac{8kT}{\pi \mu} \right)^{1/2} \tag{6.45}$$

where μ is the reduced mass of the entire Ni-CH_4 system. Expression (6.45) represents the number of collisions of a CH_4 molecule per unit of time to a surface (4.181). Comparison with the computed preexponent for methane dissociation gives an estimate of the rigidity of the transition state.

The radius r of the collision complex can be calculated from its moment of inertia. For the ratio of preexponents one finds $\nu_T/\nu_{HS} > 0.5$. The high value of the preexponent implies that the transition state is a relatively loose one, as the interaction between CH_4 and nickel results in low values for the frequencies of the NiC and NiH bonds, which reduces the corresponding partition functions relatively little, because they replace the strong methane bonds. The preexponential for methane dissociation on a surface cluster is lower, however, the ratio ν_T/ν_{HS} drops from 0.5 to 0.01.

The loose transition state for methane dissociation contrasts with the tight transition state found for the dissociation of CO in Section 6.3.3. The main

TABLE 6.4. Rate Constants for Adsorption and Desorption, Equilibrium Constant, Sticking Coefficient, and Tunneling Coefficient for Methane on Nickel[a]

Molecule	T (K)	k_{diss} ($m^3/mol \cdot s$)	k_{des} (s^{-1})	K (m^3/mol)	s	Γ
CH_4	250	$2.14 \cdot 10^0$	$6.55 \cdot 10^{-3}$	$3.27 \cdot 10^2$	$4.14 \cdot 10^{-8}$	1.486
CD_4	250	$3.38 \cdot 10^{-1}$	$3.62 \cdot 10^{-3}$	$9.34 \cdot 10^1$	$6.96 \cdot 10^{-9}$	1.419
CH_4	500	$1.09 \cdot 10^4$	$2.67 \cdot 10^5$	$4.08 \cdot 10^{-2}$	$1.49 \cdot 10^{-4}$	1.211
CD_4	500	$3.63 \cdot 10^3$	$1.81 \cdot 10^5$	$2.01 \cdot 10^{-2}$	$5.29 \cdot 10^{-5}$	1.181
CH_4	750	$2.20 \cdot 10^5$	$9.55 \cdot 10^7$	$2.30 \cdot 10^{-3}$	$2.45 \cdot 10^{-3}$	1.134
CD_4	750	$9.62 \cdot 10^4$	$6.83 \cdot 10^7$	$1.41 \cdot 10^{-3}$	$1.14 \cdot 10^{-3}$	1.115
CH_4	1000	$1.10 \cdot 10^6$	$1.82 \cdot 10^9$	$6.03 \cdot 10^{-4}$	$1.06 \cdot 10^{-2}$	1.098
CD_4	1000	$5.53 \cdot 10^5$	$1.34 \cdot 10^9$	$4.13 \cdot 10^{-4}$	$5.69 \cdot 10^{-3}$	1.083

[a]From Burghgraef *et al.* (1993).

difference is in the coordination of the dissociation products. Carbon and oxygen atoms require adsorption sites with a high coordination, i.e., three- and four-fold sites, whereas hydrogen atoms and methyl groups are less demanding in this respect and adsorb on top of single atoms. Figure 6.15 shows that the geometry of the transition state resembles that of the dissociated state, as the C–H_{act} bond is considerably stretched. The reaction can be considered an approach of the methane molecule to the nickel atom followed by stretching of the C–H bond. If the barrier for the reaction were located in the relative motion of the CH_4 molecule with respect to the nickel atom, the transition state would have been characterized as an early one. The situation of the barrier in the dissociating C–H_{act} coordinate, close to the final state, makes it a late transition state.

Molecular beam experiments in which CH_4 dissociation is studied as a function of molecular energy show a significant enhancement of the sticking coefficient when CH_4 modes are vibrationally excited. This agrees with the notion of an energy barrier for the C–H_{act} stretching mode as the reaction coordinate. The isotope effect in the reaction rate evident from the data in Table 6.3, is mainly due to differences in E_{crit} resulting from altered zero-point frequencies (6.43).

The activation energy for dissociation of methane on a nickel atom, 40.7 kJ/mole, is smaller than that for dissociation on a nickel surface. This is due to a reduction in the reactivity of a metal atom sharing bonds with neighboring atoms. The experimentally measured activation energy is 53 kJ/mole. At a temperature T = 500 K, this gives a value of the sticking coefficient according to transition state theory of about 10^{-5}. Since the experimentally observed values are 10^{-7}, it can be concluded that the trapping coefficient α of CH_4 has to be on the order of 10^{-2} (6.25). The low sticking coefficient for dissociative adsorption of CH_4 forms the reason that exposure of nickel surfaces to methane at moderate temperature in ultra high vacuum does not result in chemisorption as the number of reactive collisions of methane is too low. Elevated pressures are needed in order to enhance the collision rate. The reactivity of larger hydrocarbons is significantly higher because of a higher heat of adsorption which results in higher trapping coefficients at typical reaction temperatures, as this chapter will cover.

6.4.2. The Transition State for Methane Activation by Brønsted-Acid Protons

Solid acids like zeolites activate molecules by protonation. The resulting charge separation generates an electrostatic potential. The cost of charge separation is minimized when at least two protons of the CH_5+ carbonium remain directed to two negatively charged oxygen atoms surrounding the aluminium ion of the zeolite lattice. Again the transition state of this coordinatively saturated molecule is a loose one.

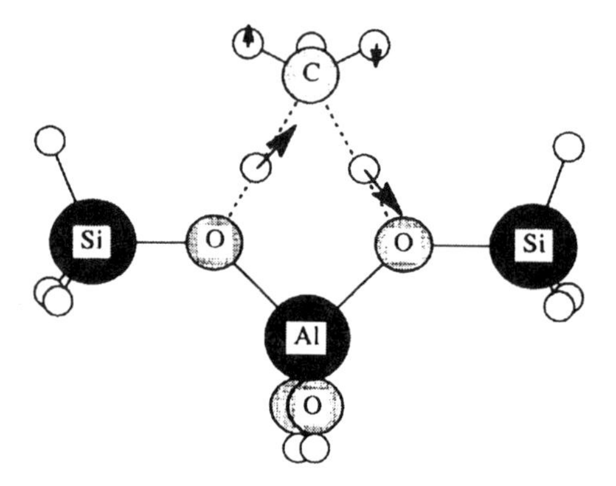

Figure 6.17. Calculated transition state geometry for H-D exchange between a proton of a zeolite and methane, CD_4; the zeolite is represented by the cluster H_3Si-O**H**-Al$(OH)_2$-O-SiH_3 (from Kramer *et al.*, 1993).

The activation of CH_4 by protons forms a prototype for the activation of C–H bonds in reactions of hydrocarbons on zeolites. As briefly introduced in Chapter 3, zeolites are microporous aluminosilicates, which contain channels of dimensions comparable to those of organic molecules. Zeolites in protonated form act as catalysts for many processes in the petrochemical industry. Figure 6.17 shows a cluster with the composition H_3Si-O**H**-Al$(OH)_2$-O-SiH_3, that mimics the catalytically active site of a zeolite very well. The aluminum ion of the cluster is tetrahedrally coordinated to oxygen atoms, just as silicon and aluminum ions in the lattice of a true zeolite are. The proton indicated in bold type is bonded to a bridging oxygen between the Si^{4+} and Al^{3+}, which gives it strong Brønsted-acidic character. Acid catalyzed reactions are initiated by elementary reaction steps as (Chapter 3)

$$ZH + CH_4 \rightarrow Z^- \cdots CH_5^+ \tag{6.46}$$

The CH_5^+ group, having a carbon atom in five-fold coordination, is called a carbonium ion. Carbonium-ion formation is an easier process in larger hydrocarbons, where the reaction is between the proton and the secondary or tertiary carbon atom, but is significantly more activated in the case of methane. The carbonium ion is the intermediate toward carbenium ions (3.29), where the positively charged carbon atom has a threefold coordination. Once formed, carbenium ions can be converted to hydrocarbons and new carbonium ions, in hydride transfer reactions, as in (3.31). Because of the lower activation energies of such reactions, the route via carbenium ions usually dominates over mechanisms involving carbonium ions. Reactions of the carbonium-ion type are terminated by a step in which a proton is transferred to the zeolite lattice.

Methane is the simplest hydrocarbon, and basic features of solid acid catalysis can be elucidated by considering carbonium ion formation from CH_4. At present, reliable quantum-chemical calculations are only feasible by using small clusters.

TABLE 6.5. Characteristic Al–O, Si–O, O–H and C–H Distances (Å) in the cluster of Figure 6.17 in the Presence of Physisorbed Methane (Ground State) and in the Transition State of H–D Exchange

	Ground State	Transition State
$Al\text{-}O_p$	1.942	1.801
Al-O	1.706	
$Si\text{-}O_p$	1.701	1.647
Si-O	1.615	
$O_p\text{-}H$	0.949	1.399
C-H	—	1.282

As an example we discuss predictions for the rate constant of the proton-deuterium exchange reaction between deuterated methane and the zeolite cluster of Figure 6.17

$$ZH + CD_4 \rightarrow ZD + CD_3H \tag{6.47}$$

Because this reaction involves the transfer of a charged proton, it is essential that the cluster is neutral. This has been achieved by saturating the terminal bonds of the silicon by hydrogen.

Reaction (6.47) starts with the physisorption of methane on the cluster, taken as the ground state. The interaction is very weak and amounts to only a few kJ/mole. Figure 6.17 shows the prediction for the transition state for H–D exchange. The reaction coordinate represents the transfer of the proton from the zeolite (i.e., from the left oxygen) to the methane molecule, and the symmetrical return of one of the deuterium atoms to the zeolite. The energy barrier for the reaction appears to be high, 150 ± 20 kJ/mol. This is due to the fact it is a primary carbon atom which is protonated; the barrier will be significantly lower for protonation of higher hydrocarbons.

Table 6.5 contains some of the computed equilibrium distances illustrative for the cluster in the ground and the transition state. Significant shortening of Si–O and Al–O bond lengths occurs when the acidic proton moves away from the oxygen. In the transition state for H–D exchange, the Al–O and Si–O bond lengths assume values in between the extremes, indicating an appreciable covalent interaction with the CD_4 molecule.

The changes in distance illustrate the operation of the bond order conservation principle: *The more the electrons of an atom become distributed over bonds to neighboring atoms, the more each of these bonds weakens.*

Once the potential energy minima of the state of physical adsorption and the saddle point of the transition state are known, vibrational frequencies can be found

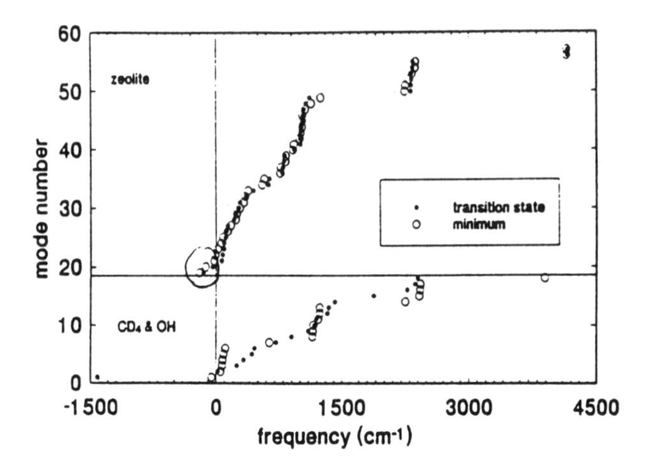

Figure 6.18. Vibrational frequencies for the adsorption minimum and transition state for the D–H exchange reaction of CD_4 on a zeolite cluster.

by expanding the potential energy in the harmonic approximation. All frequencies are displayed in Figure 6.18. The motion of the reaction coordinate, indicated by the arrows in Figure 6.17, is obtained from the imaginary frequencies associated with the curvature of the potential energy maximum in the transition state. These are indicated on the negative frequency axis of the diagram. The small imaginary frequencies with absolute values below 200 cm^{-1} are an artefact of the calculation and should be ignored, but the imaginary frequency at 1423 cm^{-1} is associated with the reaction coordinate.

The computational method gives a systematic overestimate of the frequencies by 10%. Trends in going from the ground state to the transition state have significance, however. Important shifts occur in the OH and CD_4 frequencies. The OH stretch frequency at 3890 cm^{-1} disappears in the transition state. Also the six very low frequency modes describing the weak coupling between physisorbed methane and the zeolite disappear. Instead a large imaginary frequency (*i*, 1423 cm^{-1}) corresponding to the reaction coordinate appears, together with six new modes between 20 and 710 cm^{-1} which describe the coupling between the CD_4H group and the zeolite in the transition state. The frequency of 20 cm^{-1} corresponds to the virtually free rotation of the CD_3 group in the transition state. The increased frequencies of the other modes are due to the hindered rotational and vibrational modes of the CD_5^+ complex against the deprotonated cluster.

Using these vibrational values one computes for the pre-exponent ratio $\nu_T/\nu_{HS} \approx 0.01$, which is rather similar to the value found for the activation of methane by a metal atom. The free rotation of the methyl group in the transition state is responsible for the relatively high value of ν_T. The transmission factor Γ

equals 1.4 at 500 K. Hence, the rate constant of carbonium-ion formation of methane is low mainly because of its high activation energy and not because of a small prefactor. The geometry of the transition state is indicative of a stabilization of the CH_5^+ complex by an electrostatic interaction with the negatively charged, deprotonated zeolite cluster. This interaction of CH_5^+ protons and cluster-oxygen atoms minimizes the separation of charge. As a result, the corresponding potential-energy surface is rather flat.

The activation energy for the protonation of hydrocarbons varies with the composition and the structure of the zeolite. Studies with clusters embedded in zeolite matrices show these effects are mainly due to changes in the strength of the bond between the zeolite and the acidic proton. Replacement of the hydrogen atoms that terminate the cluster of Figure 6.17, by additional tetrahedral zeolite-like groups, affects the OH bond by causing small changes in covalent interaction energies. Embedding of the cluster in an extended zeolite lattice constrains the local structure at the active site, such that the zeolite lattice resists against the geometric changes observed in the cluster calculations upon deprotonation as indicated in Table 6.5. This tends to increase the intrinsic acidity of the proton. Long-range effects associated with the different structures of zeolites derive mainly from the resulting differences in local compressibilities of the lattices.

6.5. KINETICS OF DISSOCIATIVE ADSORPTION

The overall activation energy of dissociation depends on the molecular heat of adsorption and the activation energy of dissociation for the molecule in the adsorbed state. The overall activation energy of dissociation increases with gaining coverage. This is related to a change in the order of the reaction in the reactant pressure. According to Sabatier's principle, an optimum interaction strength exists between the catalyst and the reacting intermediates for which the reaction rate is a maximum.

The effective activation energy for dissociative adsorption depends not only on the activation energies of dissociation of the adsorbed molecule but also on the equilibrium constant of molecular adsorption. Consider the case where the molecularly adsorbed species AB_{ads} is the majority reaction intermediate (mari) in the surface reaction from AB to adsorbed atoms. In order to understand the kinetics of dissociative adsorption, it is essential to realize that surface sites, indicated with *, play the role of a reactant in the kinetic equations. The scheme of reactions becomes:

$$AB + * \underset{k_{ads}}{\overset{k_{des}}{\leftrightarrows}} AB_{ads}$$

$$AB_{ads} + * \underset{k_{diss}}{\rightarrow} A_{ads} + B_{ads} \tag{6.48}$$

Note that not only the elementary rate of adsorption, but also the elementary rate of dissociation of the adsorbed molecule depends on the concentration of surface vacancies. When a molecule dissociates, positions for the dissociation products have to be available.

Because we are interested in the case that AB_{ads} is the majority reacting intermediate, implying that the rate constant for dissociation is small, the coverage of AB_{ads} is written as

$$\theta_{AB} = K[AB]\theta_* \tag{6.49}$$

The site balance becomes

$$\theta_* + \theta_{AB} + \theta_A + \theta_B = 1 \tag{6.50}$$

and the coverage of free sites is written as

$$\theta_* = \frac{1 - \theta_A - \theta_B}{1 + K[AB]} \tag{6.51}$$

At the beginning of the reaction where the coverages of A_{ads} and B_{ads} are negligible, there is

$$\theta_{AB} = \frac{K[AB]}{1 + K[AB]} \tag{6.52}$$

and the rate of dissociation becomes essentially

$$r_{diss} = k_{diss}\, \theta_{AB} \sim (1 - \theta_{AB}) \tag{6.53}$$

which, together with (6.52) equals

$$r_{diss} = \frac{k_{diss}K[AB]}{(1 + K[AB])^2} \tag{6.54}$$

Figure 6.19 illustrates this type of behavior for the rate of CO dissociation. To produce the dependence of the CO dissociation on temperature and pressure, CO was preadsorbed on a rhodium catalyst, and the rate of CH_4 production was studied by exposing the catalyst to a gas flow of hydrogen. As the reaction is rate-limited in CO dissociation, it provides a probe for the reaction rate (6.54). Note that the rate of CO dissociation has a maximum close to $\theta_{CO} = ½$, as predicted by (6.53). The analysis given above presupposes that the adsorption occurs randomly and does not

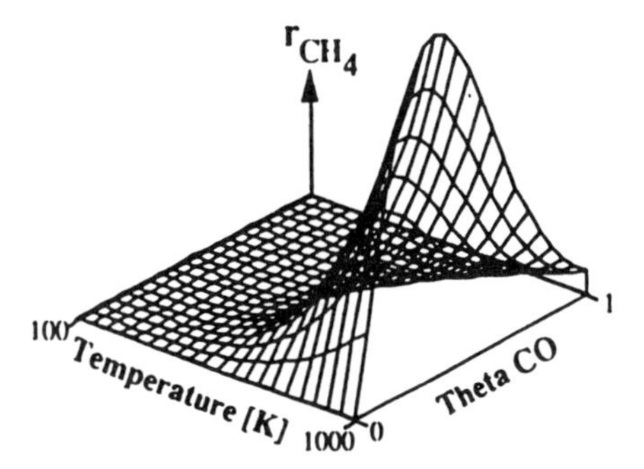

Figure 6.19. The rate of methane formation reflects the rate of CO dissociation as a function of CO coverage on the surface of a rhodium catalyst (from Koerts *et al.*, 1992).

lead to ordering/disordering phenomena. In particular, island formation of the molecular adsorbate would decrease the rate of dissociation considerably.

The overall activation energy for the dissociation reaction (6.48) becomes by definition

$$\begin{aligned} E_{act}^{eff} &= RT^2 \frac{\partial}{\partial T} \ln r_{diss} \\ &= E_{act}^{diss} + \Delta H_{ads} - 2\Delta H_{ads}\theta_{AB} \end{aligned} \tag{6.55}$$

The heat of adsorption, ΔH_{ads}, is a negative quantity. Thus, for a relatively empty surface, the overall activation energy of dissociation is an amount $|\Delta H_{ads}|$ lower than the intrinsic activation energy of the dissociation step: The heat liberated upon adsorption of the molecule AB is available for surmounting the energy barrier of the dissociation step. Figure 6.20 illustrates this with a potential energy diagram.

In contrast, under conditions where θ_{AB} is close to unity, when the pressure is high or the temperature low, the overall activation energy (6.55) is an amount $|\Delta H_{ads}|$

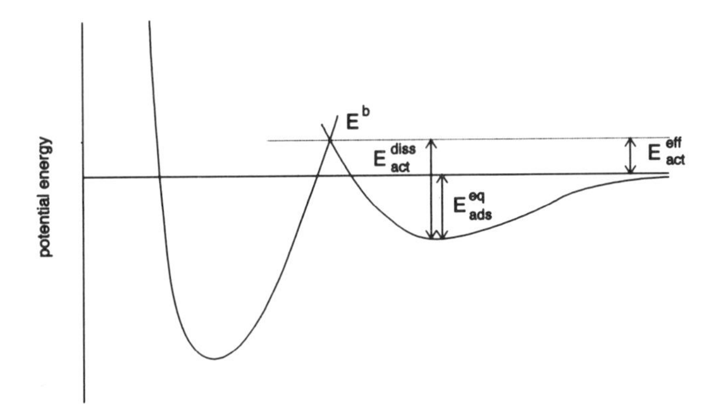

Figure 6.20. The overall activation energy for dissociation depends on surface coverage.

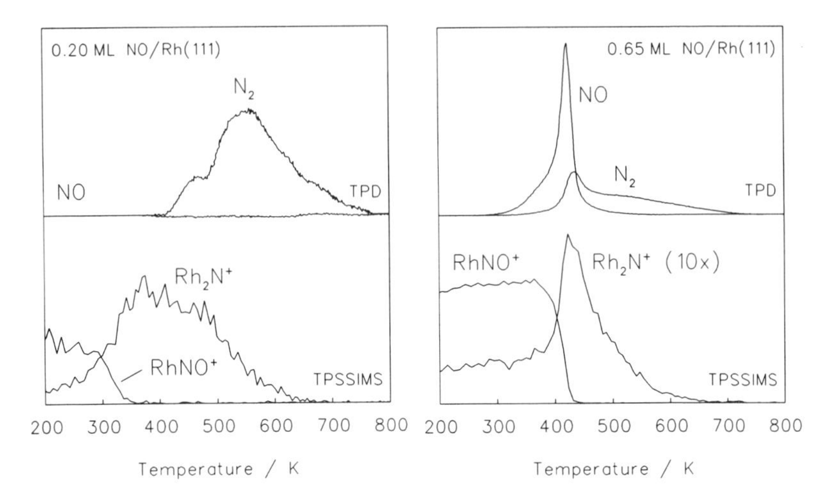

Figure 6.21. Temperature programmed SIMS and desorption experiments show that NO dissociates completely when present at low coverages (left), while dissociation at high coverages is retarded to temperatures where NO desorbs (from Borg *et al.*, 1994).

higher than the activation energy of the elementary dissociation step. The reason is there are no sites available for dissociation, so it costs an amount of energy equal to $|\Delta H_{ads}|$ to free a neighboring site. It should be noted that expression (6.55) needs to be modified for dissociations requiring more than one free site, as the factor of two in (6.55) derives from the power of the denominator in (6.54).

Temperature programmed measurements on the dissociation kinetics of NO on the (111) surface of rhodium illustrate the behavior described above. Figure 6.21 shows that NO, preadsorbed at low temperature and at a low coverage of about 15%, dissociates completely at temperatures between 250 and 350 K. This is seen in a static SIMS experiment from the disappearance of a $RhNO^+$ signal that is characteristic of molecularly adsorbed NO, and the growth of a Rh_2N^+ signal characteristic of adsorbed nitrogen atoms.† Desorption of molecular N_2 starts around 440 K. The results indicate that all NO dissociates at low coverages. If, however, the surface is saturated with NO, dissociation does not set in until a significant fraction of the molecular NO has desorbed, as the thermal desorption spectrum of Figure 6.21 shows. This retards the dissociation of NO to temperatures around 400 K, where recombination of N-atoms and desorption of N_2 follows almost instantaneously. Thus, the experiments of Figure 6.21 clearly show that if the coverage of NO is high, desorption of NO is necessary in order to create free sites for its dissociation.

---

†SIMS: Secondary ion mass spectrometry; can be used for monitoring the kinetics of surface reactions under ultra high vacuum conditions, see also Figure 2.4 (Borg and Niemantsverdriet, 1994).

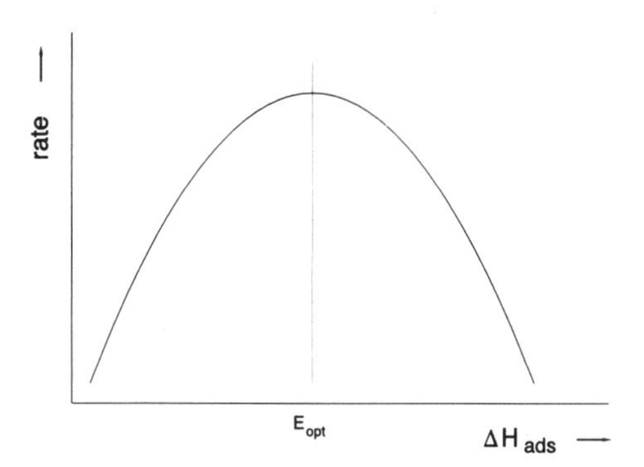

Figure 6.22. The rate of a catalytic reaction as a function of the heat of adsorption of the majority reacting intermediate.

Heats of adsorption play a key role in determining the overall activation energy of a catalytic reaction. If plotting the rate of the reaction as a function of the heat of adsorption, $|\Delta H_{ads}| = RT \ln K$, the behavior as sketched in Figure 6.22 will be found. This figure represents Sabatier's principle. Sabatier recognized that in order for a catalyst to be active, it should form a sufficiently stable surface intermediate with the reacting molecules. If the interaction is too weak, the intermediate cannot form, but if the interaction is too strong, the intermediate becomes unreactive. Thus there is an optimum interaction energy for which the rate of reaction reaches a maximum. Sabatier's principle has the kinetic implication that the order of the reaction changes as a function of increasing interaction energy. For reactants with a heat of adsorption well below the optimum, the reaction has a positive order, and the surface coverage is low. For high heats of adsorption, to the right of the optimum in Figure 6.22, the reaction has a negative order in the reactant and the surface coverage is high. Plots shown in Figure 6.22 have also been called "Volcano curves."

The heat of adsorption of a reactant can be varied by changing the composition of the catalyst. For instance, the heat of adsorption and hence the equilibrium constant decreases when one moves from the left to the right in the same row of the periodic system (Section 6.3.3). Also additives added as promoters to metal catalysts may change the adsorption energy.

Sabatier's principle has an interesting consequence for the kinetics of reactions catalyzed by systems with a wide reactivity distribution of active sites. Since the rate of reaction is at maximum for those sites having interaction energies close to the optimum of Figure 6.22, the overall rate of reaction is dominated by these sites. For this reason the kinetics of the reaction can often be modeled by equations corresponding to one type of catalytically reactive site only. However, depending

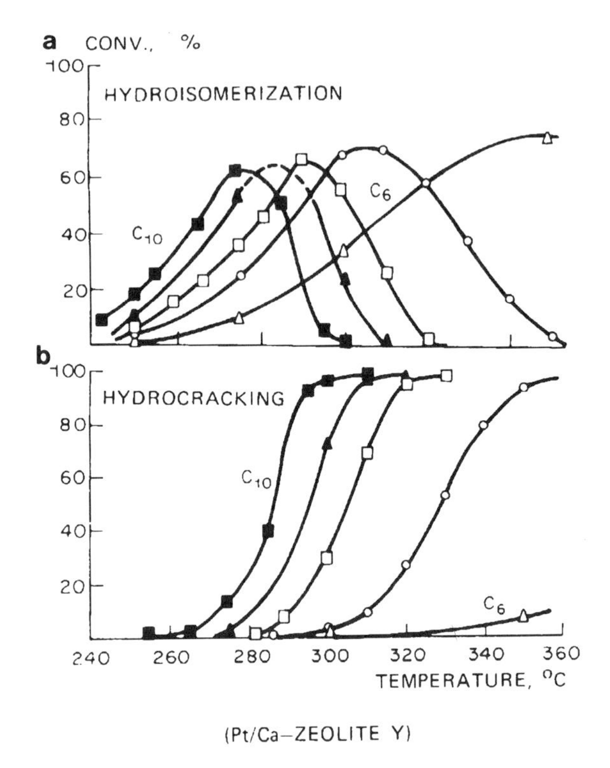

Figure 6.23. The activity of a metal/zeolite catalyst for the hydroisomerization of C_6 to C_{10} alkanes increases according to (6.58) with the chain length of the reactant (from Weitkamp, 1975).

on the conditions, the optimum interaction energy may vary and this variation will be different than for a homogeneous distribution of reactive sites.

Molecular heats of adsorption play a role in many catalytic reactions. Figure 6.23 illustrates this for an isomerization reaction catalyzed by a solid acid. As explained in Chapter 3, the hydroisomerization of alkanes on a zeolite-supported metal proceeds through a bifunctional reaction mechanism, in which the metal has the function of activating C–H bonds and H_2 at a low reaction temperature. The alkane–alkene equilibrium is established by metal catalysis, and the alkene is protonated and isomerized by the acidic protons of the zeolite

$$\begin{gathered} \text{alkane} \underset{\text{metal}}{\leftrightarrows} \text{alkene} \\ \text{alkene} \underset{\text{H}^+}{\leftrightarrows} i\text{-alkene} \\ i\text{-alkene} \underset{\text{metal}}{\leftrightarrows} i\text{-alkane} \end{gathered} \tag{6.56}$$

The isomerization is the rate-limiting step of the reaction, hence the rate is

$$r_{\text{iso}} = V\frac{d}{dt}[i\text{-alkane}] = N_{H^+}\, k_{\text{iso}}\, \theta_{alkene} \tag{6.57}$$

Similarly as in (6.48), the coverage of the alkene on the acid sites is controlled by the equilibrium constant of adsorption. Therefore, at low coverage the effective activation energy of the isomerization reaction becomes (6.55)

$$E_{\text{act}}^{\text{eff}} = E_{\text{act}}^{\text{iso}} + \Delta H_{\text{ads}} \tag{6.58}$$

The effective activation energy of the reaction becomes a function of the adsorption energy. Whereas the activation energy for isomerization to the singly branched isomer depends only weakly on the length of the hydrocarbon. The heat of adsorption increases almost linearly with the chain length. This means the effective activation energy for isomerization decreases for longer hydrocarbons. This is in accordance with the observation that the reaction temperatures needed for hydroisomerization of alkanes decreases as a function of chain length.

Figure 6.23a shows the conversion of the n-alkane as a function of temperature. In the hydrocracking reaction, the cracked products, which are formed by consecutive reactions of the isoalkane, are of interest. The maxima in the selectivity curves correspond to temperatures where cracking becomes dominant. Hydrocarbons with more than six carbon atoms can give doubly branched isomers. These molecules have an enhanced cracking rate, because β-CH scission can occur via tertiary carbenium ions (Section 3.2.3). This shifts the selectivity for cracking of longer hydrocarbons to lower temperatures.

When comparing rates measured at low or high pressures it is important to realize there may be a change in the rate-limiting step in the reaction sequence, basically as a consequence of a major change in surface coverage. Thus, if trying to relate high pressure data, where θ is high, to those obtained at low pressures where θ may be low, it is essential to construct overall rate equations based on the complete set of elementary steps measured at low pressures (Section 6.1). It appears that such procedures can be remarkably successful. They have been used to predict rates of metal-catalyzed reactions under practical conditions based on data obtained at UHV conditions.

6.6. CATALYTIC REACTIONS

6.6.1. The Hougen–Watson Approach

For catalytic reactions in which the surface reaction is the rate-determining step, the overall reaction rate is the product of three terms: the rate constant of the surface reaction, the number of unoccupied active sites, and the thermodynamic driving force.

The concept of the majority reactive intermediate (mari), as introduced in Chapter 2, is the inevitable consequence of the fact that reactants have to compete for adsorption sites on the surface of the catalyst. In general, one of the reactants or product molecules will dominate on the surface. As seen in Section 2.6, the temperature dependence of a catalytic reaction rate is closely connected with the way surface concentrations change with reaction temperature. At high temperatures, the surface sites become the majority species, and in spite of the high rate constants at elevated temperatures, the reaction rate goes down due to the lack of adsorbed reactants. Not only the activity but also the selectivity of catalyst is controlled by the concentrations of adsorbed species, as a reaction can usually follow different reaction pathways.

In this section, the emphasis is on the role of competitive adsorption phenomena in the kinetics of a few catalytic reactions. Trends in catalytic reactivity will again be discussed in terms of Sabatier's principle.

Returning to a rather general characteristic of catalytic reaction mechanisms, consider the following schematic reaction

$$A + B \underset{r^f}{\overset{r^b}{\leftrightarrows}} C + D \tag{6.59}$$

The overall rate of reaction is the sum of the forward and the backward rates

$$r = r^f - r^b \tag{6.60}$$

In Langmuir–Hinshelwood kinetics, the reaction only proceeds when reactants A and B adsorb on the surface of the catalyst. Ignoring lateral interactions and assuming the reaction is first order in the coverages of A and B, the rate of the forward reaction becomes

$$\text{Langmuir–Hinshelwood: } r^f = Nkf\,\theta_A\theta_B \tag{6.61}$$

An alternative mechanism follows from Eley–Rideal kinetics, in which reactions between adsorbed species, say A_{ads}, and molecules B in the gas phase are allowed

$$\text{Eley–Rideal: } r^f = Nkf\theta_A\,[B] \tag{6.62}$$

The practical difference between (6.62) and (6.61) is that the order in B would generally be lower in (6.61).

Competitive adsorption plays no role in the Eley–Rideal mechanism, but it is a key feature in Langmuir–Hinshelwood kinetics. In (6.59) the rate of reaction reaches its maximum when the surface species A and B are ideally mixed, and both surface concentrations are equal to one-half. Realize that the assumption of ideal mixing in the adsorbate layer is implicitly present in (6.61). The expression for the overall reaction rate becomes

$$r = Nkf\,\theta_A\theta_B - Nk^b\theta_C\theta_D \tag{6.63}$$

Assuming Langmuir adsorption and equilibrium between gas and surface phases

$$\theta_i = \frac{K_i[i]}{1 + \sum_j K_j[j]};\; i,j = A,B,C,D \tag{6.64}$$

Substitution of (6.64) into (6.63) gives the Langmuir–Hinshelwood–Hougen–Watson expression

$$r = N\,k^f \frac{K_A K_B[A][B] - K^{-1}K_C K_D[C][D]}{(1 + \sum_i K_i[i])^2} \tag{6.65}$$

Using the affinity concept of Chapter 2, (6.65) can be rewritten in the form

$$r = k^f \frac{K_A K_B[A][B]\,(1 - e^{-\bar{A}/RT})}{(1 + \sum_i K_i[i])^2} \tag{6.66}$$

in which the chemical affinity as defined in (2.12) equals

$$\bar{A} = -RT\ln\left(\frac{K_C K_D[C][D]}{K_A K_B[A][B]}K^{-1}\right) \tag{6.67}$$

If (6.66) is written in the form

$$r = kfK_A K_B[A][B]\frac{N}{(1 + \sum_i K_i[i])^2}(1 - e^{-\bar{A}/RT}) \tag{6.68}$$

then the overall rate of a catalytic reaction is seen to consist of essentially three factors: the rate of a homogeneous gas-phase reaction, the concentration of surface vacancies, and the thermodynamic driving force

$$r = k[A][B] \cdot [\text{surface vacancies}]^p \cdot \text{driving force} \tag{6.69}$$

The following looks at the importance of the different kinetic concepts in catalysis in the context of a few important catalytic reactions.

6.6.2. Competitive Adsorption in Hydrocarbon Catalysis

In hydrocarbon conversion on transition metal catalysts, the rate of reaction is largely controlled by competitive adsorption of hydrogen and the reacting hydrocarbon.

The following example of hydrocarbon catalysis illustrates the consequences of competitive adsorption for the order of the reaction, and the optimum conditions under which the reaction is carried out. The selectivity of the transition-metal catalyzed conversion of n-hexane to i-hexane versus the hydrogenolysis to smaller hydrocarbons depends strongly on the structure of the surface. Hydrogenolysis requires an ensemble of several atoms. When sites become blocked by inert atoms, such as carbon or sulfur, the selectivity for reactions maintaining chain length is significantly enhanced.

The rates of hydrocarbon conversion reactions are strongly dependent on the partial pressure ratio of hydrogen and the hydrocarbon, as illustrated in Figure 6.24 for the ring-opening reaction of methyl cyclopentane. Hydrogen is needed for product formation. In the absence of hydrogen, methyl cyclopentane adsorbs and becomes dehydrogenated to some carbonaceous residue. Although the C–H bond of a hydrocarbon is stronger than a C–C bond, the former is the first to be activated by a metal surface. Only when C–H bonds have been broken, a direct interaction

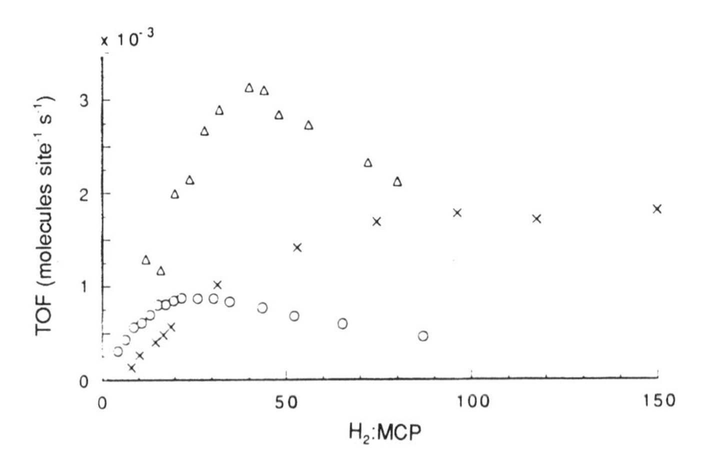

Figure 6.24. Activity for the ring opening of methyl cyclopentane as a function of $[H_2]/[MCP]$ ratio for platinum supported on (Δ) a zeolite and (*o*) on Al_2O_3 (from Vaarkamp, 1993).

between molecular carbon atoms and metal surface atoms becomes possible which leads to the cleavage of a C–C bond.

Hydrogenolysis requires the addition of two hydrogens to the molecular fragments in order to saturate the broken C–C bond. The rate follows from an expression of the type

$$r_{\mathrm{H}} = -V\frac{d[\mathrm{MCP}]}{dt} = N\,k_{\mathrm{H}} \cdot \theta_{\mathrm{H}}^{2} \cdot \theta_{\mathrm{MCP}} \tag{6.70}$$

where MCP stands for methyl cyclopentane. For low ratios of the H_2 to hydrocarbon partial pressure, one finds for catalysis by platinum that the rate has a positive order (between 0.5 and 1, depending on the type of support used for the platinum catalyst) in the hydrogen pressure, as well as in the methyl cylopentane as expressed in the following rate expression

$$r_{\mathrm{H}} = -V\frac{d[\mathrm{MCP}]}{dt} = N\,k_{\mathrm{H}}K_{\mathrm{H_2}}K_{\mathrm{MCP}}\,[\mathrm{H_2}]^2[\mathrm{MCP}]\,\theta_*^3 \tag{6.71}$$

with

$$\theta_* = \frac{1}{1 + \sqrt{K_{\mathrm{H_2}}[\mathrm{H_2}]} + \mathrm{K_{MCP}}[\mathrm{MCP}] + \ldots\ldots} \tag{6.72}$$

The term $K^{1/2}[H_2]^{1/2}$ in the denominator of (6.72) is insignificant for low partial pressures of hydrogen, causing the positive order in hydrogen. At higher hydrogen pressures, corresponding to $[H_2]/[MCP]$ of about 20–50, the rate of the ring opening goes through a maximum until at very high ratios the rate obtains a negative order in hydrogen (between –0.6 and –1). Here adsorbed hydrogen atoms form the majority reacting intermediate and the term $K^{1/2}[H_2]^{1/2}$ becomes dominant in the denominator of the expression for θ_*. The reaction is actually suppressed by hydrogen. Figure 6.24 and the expressions (6.71) and (6.72) illustrate that the catalytic behavior of platinum particles in the ring opening of methyl cyclopentane can largely be explained in terms of competitive adsorption of the reactants.

The local environment of the metal particles in the zeolite can be varied by changing the composition of the zeolite. In zeolite L, the metal particles are located at close distances of channel cations that compensate for the negative charge of the zeolite lattice. Systematic studies of the hydrogenation of benzene as well as the aromatization of n-hexane have been performed as a function of the size of alkaline-earth cations in the zeolite. Figure 6.25 illustrates the dependence found.

The presence of a cation close to a small metal particle polarizes the latter. The geometry of adsorption is illustrated in Figure 6.26. Aromatic molecules become slightly electron deficient upon adsorption. This implies that the withdrawal of charge from the surface of the metal particle toward the cation underneath is beneficial for the adsorption of benzene. Benzene hydrogenation rate is enhanced

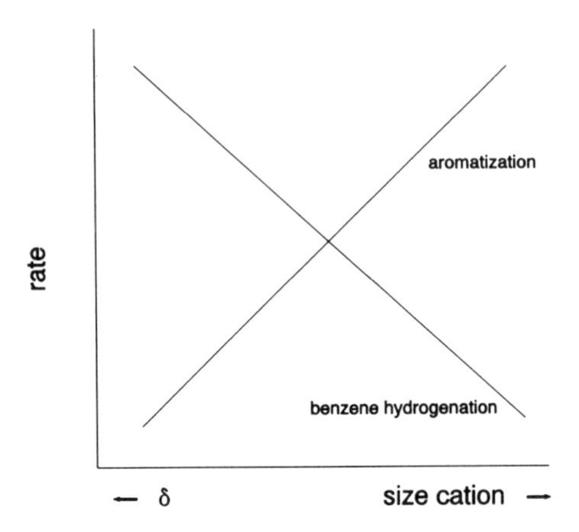

Figure 6.25. The rate of hydrocarbon conversion as a function of alkaline-earth cation radius (schematic).

by the increased value of the equilibrium constant of the adsorption. As small cations polarize the metal particle most, benzene hydrogenation rate is high when the zeolite contains small alkaline-earth cations, as indicated schematically in Figure 6.25. The aromatization of n-hexane to benzene shows the reverse trend, as the rate-limiting step of this reaction is the desorption of benzene. In this case, the

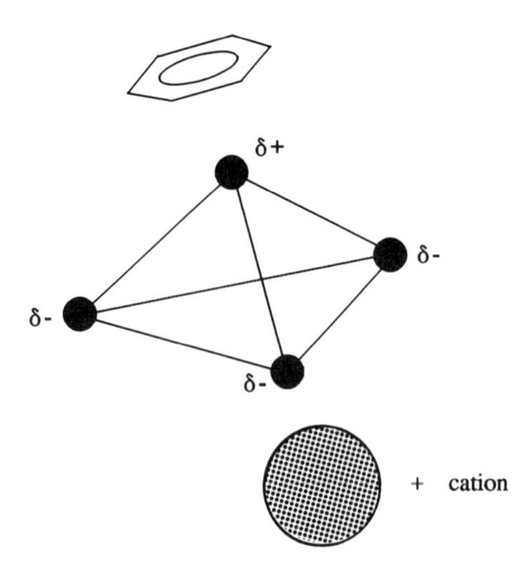

Figure 6.26. Polarization of a metal particle in the field of a cation.

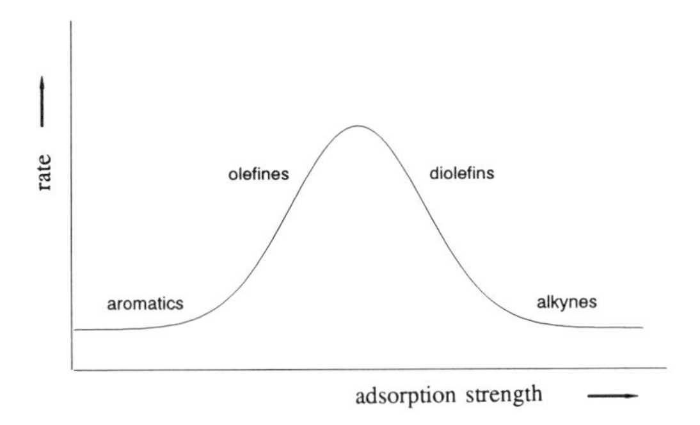

Figure 6.27. Schematic graph of activity in hydrogenation reactions as a function of the adsorption strength on palladium (after Hub *et al.*, 1988).

reaction is favored by a weaker interaction between the aromatic molecule and the metal, as is promoted by a cation of larger size.

In Section 6.5 it was argued the effective activation energy of a dissociation reaction depends not only on the elementary step that is rate-determining, but also on the heat of adsorption of the reactant. At low coverages it lowers the effective activation energy. When the heat of adsorption is too strong, however, the surface coverage increases and the reaction obtains a negative order in the pressure of the corresponding reactant and the effective activation energy increases. Figure 6.27 illustrates this principle for the hydrogenation of several types of hydrocarbons.

Alkynes with their triple carbon–carbon bond adsorb strongly on the surface of transition metals, causing the hydrogenation rate to be low. Aromatics, on the other hand, adsorb weakly, with the consequence that the hydrogenation rate is low due to a low surface coverage of the reactant. Olefin hydrogenation comes close to the optimum interaction energy causing the hydrogenation rate to be high, as expected on the basis of Sabatier's principle.

The selectivity toward product molecules with the same number of carbon atoms as in the reactant varies significantly for the different transition metals. The stronger the metal–carbon surface bond strength, the lower the activation energy for dissociation and the lower the selectivity becomes. For this reason platinum, which represents the transition metal with the lowest reactivity, is the preferred metal for the catalytic isomerization or aromatization of alkanes. The Group I-B metals, Cu, Ag, and Au, form weaker metal–carbon bonds, but have a reactivity that is too low. In the case of Ag and Au the interaction energies become so weak that neither H_2 nor C–H bonds dissociate. On Cu such bond dissociation is possible, but with a considerable activation energy.

Platinum appears to be the metal offering the optimum interaction energies for adsorption and catalysis of hydrocarbons. The other transition metals interact more strongly with hydrocarbons, leading to higher rates for hydrogenolysis and consequently lower selectivities for reactions such as isomerization, aromatization, and hydrogenation, i.e., reactions in which the number of carbon atoms in a molecule is conserved.

6.6.3. Volcano Curve for the Synthesis of Ammonia

The rate of the ammonia synthesis shows an optimum for metals in the first column of the Group VIII metals and provides a classical example of Sabatier's principle.

The principle of Sabatier is nicely illustrated by the rate of ammonia synthesis on different metals (Figure 6.28). As explained in Chapter 3, the reaction is controlled by a competition between dissociative nitrogen adsorption and the rather strong adsorption of ammonia. The rupture of the strong N–N bond of N_2 requires a catalyst that interacts strongly with the nitrogen atoms produced upon dissociation. Iron and ruthenium appear as suitable catalysts located near the optimum of the Volcano curve of Figure 6.28. To the right of ruthenium in the periodic system, the metal–adsorbate interaction is too weak and the nitrogen dissociation rate is expected to be rate limiting. On the contrary, to the left, nitrogen atoms form strong bonds with the catalyst, causing the activation energy for the surface reaction to be high. The rate expression given by (3.7) has the Langmuir–Hinshelwood–Hougen–Watson form, showing the rate is indeed suppressed by adsorption of ammonia.

The most commonly used ammonia synthesis catalyst is iron promoted by potassium. The alkali promoter facilitates the dissociation of nitrogen and weakens the interaction of ammonia with the surface of the catalyst. The latter is readily understood if it is realized that the basic ammonia binds to iron by donation of its lone-pair electrons to the metal. This donation becomes less favorable in the

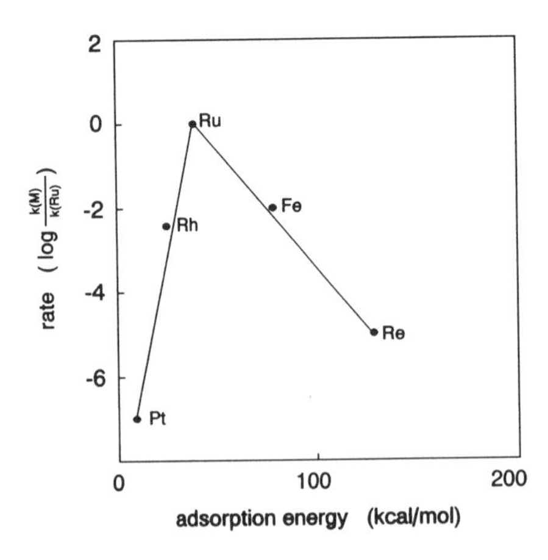

Figure 6.28. Volcano curve for the rate of ammonia synthesis on different metals (from Jennings, 1991).

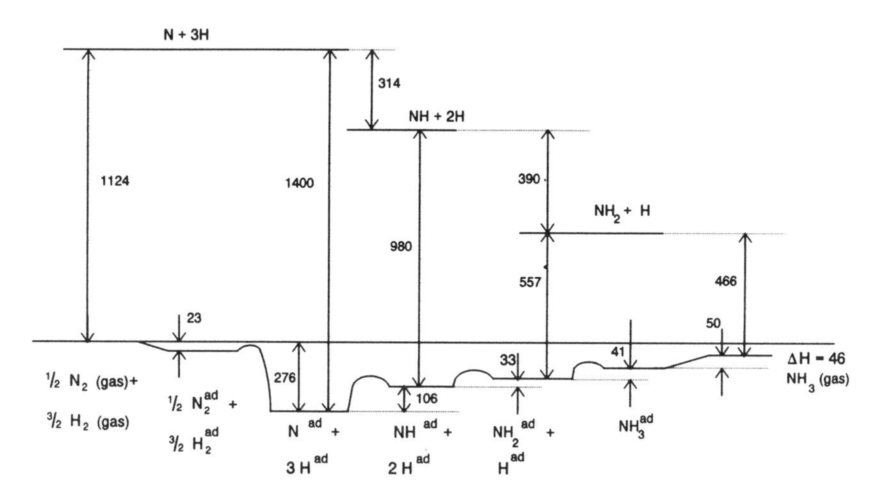

Figure 6.29. Schematic energy profile for the ammonia synthesis on a promoted iron catalyst, with energies in kJ/mol (from Ertl, 1983).

presence of positively charged alkali stabilizing the electron energies of ammonia, with respect to the metal surface.

Investigations in surface science by Ertl and coworkers have revealed many details of the elementary steps involved in the ammonia synthesis. Figure 6.29 shows the energy profile. The interaction between nitrogen and iron is such that the dissociation of the nitrogen molecule is not activated with respect to the gas phase. The adsorbed nitrogen atom becomes hydrogenated by atomic hydrogen in consecutive reaction steps. According to the bond order conservation principle, the metal-nitrogen bond weakens with each subsequent hydrogen addition, therefore, the reaction between a bare nitrogen and hydrogen is the most difficult of the three hydrogenation steps.

6.6.4. Selectivity Control by Competitive Adsorption

The selectivity of recombination reactions on a surface is controlled by the composition of the adsorbate layer. The concentrations of adsorbed species may vary with the reactivity of the catalytic surface, and so does the selectivity. Similar principles apply in homogeneous reaction mechanisms.

Competitive adsorption phenomena are often responsible for selectivity patterns in catalytic reactions. An interesting case is found in the field of automotive emission catalysis, and concerns the reduction of NO in a gas mixture with CO and

H_2 on noble metal catalysts, platinum, and rhodium. Since one of the antibonding orbitals of NO bond is occupied by an electron, the N–O bond is weaker than the bond of CO. The rate of dissociative NO adsorption is generally faster than that of CO. Rhodium is more reactive than platinum. Therefore the rate of dissociative adsorption of NO is higher on Rh than on Pt, as is the steady-state coverage of adsorbed nitrogen atoms.

The desired overall reaction is

$$NO + CO \rightarrow \frac{1}{2}N_2 + CO_2 \tag{6.73}$$

However, for a surface covered with N, H, O atoms, and CO molecules, several other reactions are conceivable. The production of ammonia according to

$$N_{ads} + 3H_{ads} \rightarrow NH_3\uparrow \tag{6.74}$$

is certainly undesirable and competes with other surface reactions such as

$$N_{ads} + N_{ads} \rightarrow N_2\uparrow \tag{6.75}$$

and

$$N_{ads} + NO_{ads} \rightarrow N_2\uparrow + O_{ads} \tag{6.76}$$

while CO_2 is formed by recombination of O_{ads} with CO_{ads}

$$CO_{ads} + O_{ads} \rightarrow CO_2\uparrow \tag{6.77}$$

Clearly, a high surface coverage of N_{ads} favors the reaction to N_2. At low coverages of N_{ads} the chance to recombine with other adsorbed atoms like H_{ads} increases. According to Nieuwenhuys this competitive adsorption mechanism explains the higher selectivity of platinum for NH_3 as compared to rhodium. The surface concentration of N_{ads} is significantly lower and the coverage of H_{ads} is higher than on rhodium, causing a higher probability for reaction (6.74).

Selectivity control by competitive adsorption also explains how the platinum catalyst works in the Ostwald process, the oxidation of ammonia to nitrogen oxide, used to produce nitrates from ammonia. The low reactivity of platinum causes the rate of ammonia dissociation on platinum to be low. The dissociation of N–H bonds of ammonia is the reverse of the ammonia synthesis discussed in the previous section. Whereas the rate of dissociative NH_3 adsorption is relatively small, the initial sticking coefficient for dissociative oxygen adsorption is higher, about 0.01.

Two competing reactions exist

$$N_{ads} + O_{ads} \rightarrow NO\uparrow$$

$$N_{ads} + N_{ads} \rightarrow N_2\uparrow \tag{6.78}$$

Thermodynamically, the formation of N_2 is preferred, but because the steady-state concentration of N_{ads} is low and that of O_{ads} relatively high, the selectivity towards NO is high. NO formation is thus kinetically controlled, and possible because of competitive adsorption in which the coverage of O_{ads} largely exceeds that of N_{ads}.

Also the field of homogeneous catalysis provides examples in which competitive adsorption controls the selectivity of a reaction. An interesting new polymer is polyketon

$$-\underset{\underset{O}{\|}}{C}-CH_2-CH_2-\underset{\underset{O}{\|}}{C}-CH_2-CH_2-\underset{\underset{O}{\|}}{C}- \tag{6.79}$$

This polymer is made by a reaction of CO and C_2H_4 by means of a catalyst consisting of a Pd^{2+} complex with a bidentate phosphine compound as a ligand. In the absence of CO, the catalyst produces polyethylene

$$-CH_2-CH_2-CH_2-CH_2-CH_2-CH_2- \tag{6.80}$$

Interaction of ethylene with the Pd^{2+} complex appears to be weak. The ligands around the Pd^{2+} ion are quite bulky and restrict the approach of ethylene to the Pd^{2+} center of the catalytically active organometallic complex. The smaller CO molecule does not experience steric hindrance and interacts strongly with the complex. As a consequence, the reaction has a positive order in ethylene, but a negative order in CO. Figure 6.30 schematically shows the Pd-complex and the growing polyketon chain after an ethylene insertion step.

After CO insertion, CO and C_2H_4 compete for adsorption on the Pd-center. Although the adsorption equilibrium clearly favors the adsorption of CO, this does not lead to insertion in the polymer, as this reaction step, sketched in Figure 6.31a appears to be strongly endothermic. The ethylene insertion step (Figure 6.31b), however, is thermodynamically favored. Thus, selective polyketon formation by

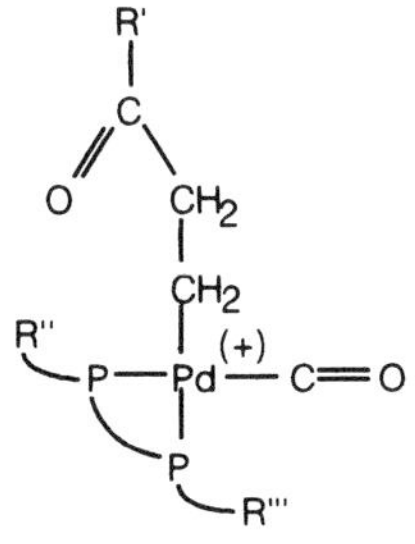

Figure 6.30. Growing polyketon chain on a palladium complex showing the CO insertion step.

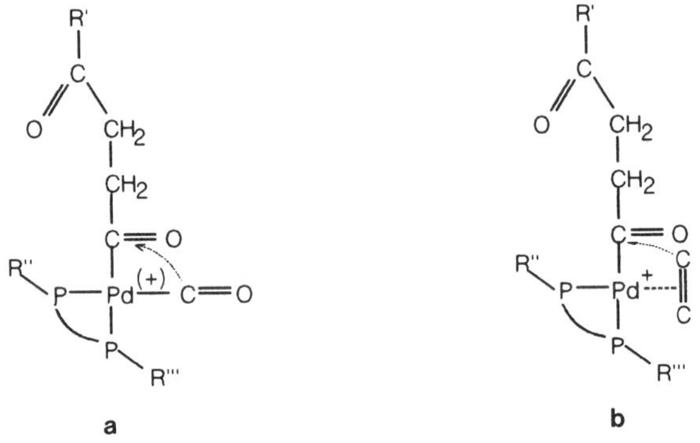

a

b

Figure 6.31. Insertion of CO in a Pd–CO bond (a) is thermodynamically unfavorable as compared to the insertion of ethylene (b).

alternating insertion of CO and C_2H_4 is possible for two reasons: Competitive adsorption between CO and C_2H_4 molecules is highly in favor of CO, reducing the rate of consecutive insertions of ethylene significantly, and insertion of a CO molecule in the metal–CO bond of the polyketon–Pd adduct is thermodynamically unlikely.

6.6.5. Selectivity in the Fischer–Tropsch Synthesis

The selectivity of the Fischer–Tropsch synthesis—the reaction of CO and H_2 to hydrocarbons—depends on the rate of dissociative CO adsorption, and on the rates of chain propagation and chain termination. Methanation, carburization, and graphitization, which leads to the deactivation of the catalyst, are competing processes. The rates of these processes are controlled by the surface concentration of hydrogen. Chain propagation and termination are also competing reactions regulated by the coverages of CH_x species (favorable for propagation) and hydrogen (favorable for termination).

The Fischer–Tropsch synthesis, i.e., the formation of hydrocarbons from CO and H_2, represents an oligomerization reaction in heterogeneous catalysis.

$$CO + H_2 \rightarrow C_nH_{2n+2}, C_nH_{2n} \tag{6.81}$$

As explained in Chapter 3, iron, cobalt, and ruthenium are successful catalysts for this process. Nickel has a high selectivity for the formation of methane, while copper and palladium favor the reaction to methanol. On rhodium, ethanol and aldehydes are formed in addition to methane and methanol. Comparison of the differences in selectivity between the different metals provides an interesting illustration for activity trends in CO dissociation and carbon–carbon bond formation as a function of metal–adsorbate interaction strength.

The reaction proceeds via dissociation of CO and H_2, generally the order of the reaction is slightly negative in CO but positive in H_2. The metal–hydride bond has a strength on the order of 250 kJ/mol, resulting in a heat of dissociative adsorption of some 70 kJ/mol. The heat of CO adsorption varies between 85–170 kJ/mol. Evidently, hydrogen and CO adsorb competitively, where CO tends to have the higher coverage. Among the Group VIII metals, platinum and iridium form the strongest bonds with CO, which is associated with the high work function of these metals. The latter favors interactions with adsorbates with electron-donating capabilities. In the case of CO, the 5σ HOMO, directed towards the metal, provides such an interaction. Platinum and iridium are poor Fischer–Tropsch catalysts, because CO dissociation is unlikely due to the high work function of the metals and the relatively weak metal–carbon and metal–oxygen bonds to be formed upon dissociation. Both factors contribute to a high activation energy for CO dissociation. In addition, the rather strong metal–CO interaction results in a high surface coverage of molecular CO, which again does not encourage dissociation. Carrying out the CO + H_2 reaction at a higher temperature leads to the production of methane as the major product. This process is accompanied by extensive graphite formation on the surface, which can be considered as the consequence of Pt–C and Ir–C bonds that are weaker than in other Group VIII metals.

The bonding of carbon and oxygen atoms to metals such as iron, cobalt, nickel, and ruthenium is significantly stronger and hence the dissociation of CO requires a much lower activation energy on these metals. Subsequent reactions with hydrogen atoms gives rise to CH_x fragments on the surface, which on nickel hydrogenate rapidly to CH_4, but on the other metals polymerize to longer hydrocarbons.

Copper is the preferred catalyst for methanol formation from synthesis gas. Carbon monoxide forms a weak bond with the copper surface. The repulsive interaction between the doubly occupied 5σ orbital of CO and the fully occupied d-band pushes CO to the on-top position, because the repulsion is minimized for adsorption geometries of low coordination. In this position the $2\pi^*$ bond of CO is barely activated and dissociation is not possible. Methanol is thought to be formed through formate ($HCOO_{ads}$) intermediates. Such species are more readily produced from CO_2 and H_2, which is the feed in the practical methanol synthesis. Starting from CO, formate can be formed by a reaction with a hydroxyl group. The formate is next hydrogenated to a methoxy (H_3CO_{ads}) species, which combines with an adsorbed hydrogen atom to methanol.

On rhodium, the activation energy of CO dissociation is comparable to that of CO desorption. However, the rate of dissociation is low and the surface coverage with CH_x intermediates that are necessary for chain growth is small. The simultaneous presence of CO and CH_x species on the rhodium surface enables the following insertion reaction

$$CH_3\text{–}Rh\text{–}C{=}O \longrightarrow CH_3\text{–}C({=}O)\text{–}Rh \tag{6.82}$$

which has to compete with the methanation and chain growth reactions

$$\underset{|}{CH_3} + \underset{|}{CH} \xrightarrow{H} CH_3\text{–}\underset{|}{CH_2} \tag{6.83}$$

or

$$\underset{|}{CH_3} + \underset{|}{CH_2} \longrightarrow CH_3\text{–}\underset{|}{CH_2} \tag{6.84}$$

Interestingly, reactions of (6.82) are also known in transition metal complexes of Rh^{n+} and CO.

Changing the selectivity to desirable products such as ethanol requires the addition of promoters that enhance the rate of CO dissociation just enough to give the optimum surface coverages of CO_{ads} and $CH_{x,ads}$. The chain growth reactions (6.83) and (6.84) benefit from stronger metal–carbon bonds, as this suppresses both methanation and the termination of the growing hydrocarbon, since both involve the breaking of the M–C bond. For this reason, iron, cobalt, and ruthenium, which all form relatively strong metal-carbon bonds, are excellent catalysts for chain growth.

The overall reaction scheme of the Fischer—Tropsch reaction is

$$\begin{array}{ccccccccccc} CO, H_2 & & & & & & & & & & \\ \Updownarrow & & & & & & & & & & \\ CO_{ads}, H_{ads} & \xrightarrow{r_i} & C_1 & \xrightarrow{r_p} & C_2 & \xrightarrow{r_p} & C_3 & \longrightarrow & \cdots & \xrightarrow{r_p} & C_n \\ & & \downarrow r_m & & \downarrow r_t & & \downarrow r_t & & & & \downarrow r_t \\ & & CH_4 & & C_2H_6 & & C_3H_8 & & & & C_nH_{2n+2} \end{array} \tag{6.85}$$

When the growing hydrocarbon contains more than two carbon atoms, the rate of propagation r_p and the rate of termination r_t become independent of chain length. The propagation rate r_p depends on the surface concentrations of the CH_x intermediates and of hydrogen. Each molecular fragment C_n has the choice to desorb as an alkene or to be hydrogenated to an alkane. The overall rate of the reaction based on scheme (6.85) becomes

$$\frac{dC_n}{dt} = r_t \left(\frac{r_p}{r_p + r_t} \right)^{n-1} \frac{r_i}{r_p + r_t} \theta_{CO} \theta_H^x$$

$$= \frac{r_t r_i}{r_p} \alpha^n \theta_{CO} \theta_H^x \tag{6.86}$$

with

$$\alpha = \frac{r_p}{r_p + r_t} < 1 \tag{6.87}$$

The rate of initiation, r_i, depends on the CO dissociation step and the subsequent formation of the active intermediate C_1 that is either hydrogenated to methane or inserts in the growing chain. The total rate of conversion follows from (6.86) by summation of the individual rates

$$-\frac{d[CO]}{dt} = \sum_i \frac{d[C_i]}{dt} = \frac{\alpha}{1-\alpha} \frac{r_t r_i}{r_p} \theta_{CO} \theta_H^x \tag{6.88}$$

Remember that all parameters in (6.88) depend on the partial pressures of H_2 and CO. Chain growth is controlled by the relative rates of propagation and termination. An increase in the metal–carbon interaction strength causes a higher surface coverage with C_1 intermediates and favors the propagation reaction. However, when the metal–carbon interaction becomes too strong, the rate of termination, r_t, becomes so slow that the overall reaction rate is suppressed.

Iron is the industrial Fischer–Tropsch catalyst and is applied in practice. Reduced iron interacts strongly with carbon. Because the activation energy for carbon diffusion into the metallic iron lattice is low (40–65 kJ/mol), the metal converts to iron carbides during reaction. According to Niemantsverdriet *et al.*, the initial rate of the Fischer–Tropsch reaction is low because the carburization process consumes most of the carbon. When the iron particles become saturated, carbon stays at the surface where it is available for the actual Fischer–Tropsch reaction. Molybdenum is also converted to a carbide or an oxide when exposed to synthesis gas; in this state it is an active catalyst. However, the early transition metals form stable but unreactive compounds in synthesis gas and are inactive as Fischer–Tropsch catalysts.

Thus, moving upward from right to left through the metals in Group VIII, the metal–CO bond weakens while the metal-carbon becomes stronger; consequently, the activation energy for dissociation goes down. The rate-limiting step for hydrocarbon formation changes from dissociation of CO (Pt, Ir, Rh), to the formation of CH_x species (Ni, Co, Fe) to the termination of the growing hydrocarbon chain (Ru).

Again, Sabatier's principle operates: The rate-limiting step of the reaction changes along with a variation of the interaction energy between the adsorbed species and the catalyst. This has large consequences for the selectivity of the Fischer–Tropsch reaction.

REFERENCES

C.T. Au and M.W. Roberts, *Nature* **319**, 206 (1986).
H. van Bekkum, E.M. Flanigan, and J.C. Jansen (eds.), *Introduction to Zeolite Science and Practice*, Elsevier, Amsterdam, (1991).
B.P. Belousov, *Ref. Radiats. Med.* **1**, 145 (1958).
B.E. Bent, Ph.D. Thesis, University of California, Berkeley, (1986).
P. Biloen and W.M.H. Sachtler, *Adv. Catal.* **30**, *165* (1981).
G.K. Boreskov, in *Catalysis Science and Technology* (J.R. Anderson and M. Boudart, eds.), Vol. 3, p. 39, Springer, Berlin, (1982).
H.J. Borg and J.W. Niemantsverdriet, in *Catalysis, Specialist Periodical Reports* (J.J. Spivey and A.K. Agarwal, eds.), Vol. 11, p. 1, The Royal Society of Chemistry, Cambridge, (1994).
H.J. Borg, J.F.C.J.M. Reijerse, R.A. van Santen, and J.W. Niemantsverdriet, *J. Chem. Phys.* **101**, 10052 (1994).
H. Burghgraef, A.P.J. Jansen, and R.A. van Santen, *J. Chem. Phys.* **98**, 8810 (1993).
H. Brune, J. Wintterlin, R.J. Belin, and G. Ertl, *Phys. Rev. Lett.* **68**, 624 (1992).
C.T. Campbell, Y.K. Sun, and W.H. Weinberg, *Chem. Phys. Lett.* **179**, 53 (1991).
C.D. Chang, *Catal. Rev.-Sci. Eng.* **25**, 1 (1983).
H.L. Chen and C.B. Moore, *J. Chem. Phys.* **54**, 4072 (1971).
P.D. Cobden, J. Siera, and B.E. Nieuwenhuys, *J. Vac. Sci. Technol.* **A 10**, 2487 (1992).
H. Conrad, G. Ertl, J. Koch, and E.E. Latta, *Surface Sci.* **43**, 462 (1974).
R.D. Cortright, S.A. Goddard, J.E. Resoske, and J.A. Dumesic, *J. Catal.* **127**, 342 (1991).
J.R. Creighton and J.M. White, *Surface Sci.* **129**, 327 (1983).
J.R. Creighton, K.M. Ogle, and J.M. White, *Surface Sci.* **138**, L137 (1984).
M.A. van Daelen, Y.S. Li, J. Newsam, and R.A. van Santen, *Chem. Phys. Lett.* **226**, 100 (1994).
B.H. Davis and W.P. Hettinger Jr. (eds.), *Heterogeneous Catalysis, Selected American Histories*, ACS Symposium Series 222, American Chemical Society, Washington, 1983.
K.B. Dobbs and D.J. Doren, *J. Chem. Phys.* **97**, 3722 (1992).
M. Eiswirth, P. Möller, K. Wetzl, R. Imbihl, and G. Ertl, *J. Chem. Phys.* **90**, 510 (1989).
T. Engel and G. Ertl, *Chem. Phys.* **69**, 1267 (1978).
G. Ertl, *Adv. Catal.* **37**, 213 (1990).
G. Ertl, in *Catalysis, Science and Technology* (J.R. Anderson and M. Boudart, eds.), Vol. 4, p. 273, Springer, Berlin, (1983).
J.D. Evanseck, J.F. Blake, and W.L. Jorgensen, *J. Am. Chem. Soc.* **107**, 154 (1985).
R.J. Field, E. Körös, and R.M. Noyes, *J. Am. Chem. Soc.* **94**, 8649 (1972).

W.H. Flygare, *Acc. Chem. Res.* **1**, 121 (1968).
W.S. Gallaway and E.F. Barker, *J. Chem. Phys.* **10**, 88 (1942).
B.C. Gates, J.R. Katzer, and G.C.A. Schuit, *Chemistry of Catalytic Processes*, McGraw-Hill, New York, (1979).
R. Gelten, M. Sc. Thesis, Eindhoven University of Technology, 1994.
R.G. Gilbert and S.C. Smith, *Theory of Unimolecular and Recombination Reactions*, Blackwell Scientific Publications, Oxford, (1990).
S. Glasstone, K.J. Laidler, and H. Eyring, *The Theory of Rate Processes*, McGraw-Hill, New York, (1941).
J. Gleick, *Chaos: Making a New Science*, Viking, New York, (1987); in Dutch: *Chaos: De Derde Wetenschappelijke Revolutie*, Uitgeverij Contact, Amsterdam, 4e druk, (1991).
W. Göpel, G. Rocher, and R. Feierabend, *Phys. Rev.* **B28**, 3427 (1983).
H. Heinemann, in *Catalysis. Science and Technology* (J.R. Anderson and M. Boudart, eds.), Vol. 1, p. 1, Springer, Berlin, (1981).
V.E. Henrich and P.A. Cox, *The Surface Science of Metal Oxides*, Cambridge University Press, Cambridge, (1993).
D.R. Herschbach, H.S. Johnston, K.S. Pitzer, and R.E. Powell, *J. Chem. Phys.* **25**, 736 (1956).
T.L. Hill, *An Introduction to Statistical Thermodynamics*, Addison-Wesley, Reading, Pennsylvania, (1960).
S. Holloway and G.R. Darling, *Comments At. Mol. Phys.* **27**, 341 (1992).
J. Horiuti and M. Polanyi, *Trans. Faraday. Soc.* **30**, 1164 (1934).
S. Hub, L. Milaire, and R. Touronde, *Appl. Catal.* **63**, 307 (1988).
A.P.J. Jansen, *J. Chem. Phys.* **97**, 5205 (1992).
J.R. Jennings (ed.), *Catalytic Ammonia Synthesis Fundamentals and Practice*, Plenum, New York, (1991).
H.S. Johnston, *Gas Phase Reaction Theory*, The Ronald Press Co., New York, (1966).
T. Koerts, W.J.J. Welters, and R.A. van Santen, *J. Catal.* **134**, 1 (1992).
G.H. Kohlmaier in *Chemische Elementarprozesse*, (H. Hartman, ed.), p. 139, Springer, Berlin, (1968).
A. de Koster and R.A. van Santen, *Surface Sci.* **233**, 366 (1990).
G.J. Kramer, R.A. van Santen, C.A. Emeis, and A. Novak, *Nature* **363**, 529 (1993).
H.A. Kramers, *Physica* **7**, 284 (1940).
H.H. Kung, *Transition Metal Oxides: Surface Chemistry and Catalysis*, Elsevier, Amsterdam, (1989).
K.J. Laidler, *Chemical Kinetics* (3rd ed.), Harper and Row, New York (1987).
H.C. Lee and J.B. Butt, *J. Catal.* **49**, 320 (1977).
R.D. Levine, *Adv. Chem. Phys.* **47**, 239 (1981).
R.A. Marcus, *Disc. Farad. Soc.* **29**, 21 (1960).
P. Mars and D.W. van Krevelen, *Chem. Eng Sci.* **3**, 41 (1954).
I.E. Maxwell and W.H.J. Stork, in *Introduction to Zeolite Science and Practice* (H. van Bekkum, E.M. Flanigan, and J.C. Jansen, eds.), p. 571, Elsevier, Amsterdam, (1991).
A. Mittasch, *Geschichte der Ammoniaksvnthese*, Verlag Chemie, Weinheim, (1951).
J.A. Moulijn, P.W.N.M. van Leeuwen, and R.A. van Santen (eds.), *An Integrated Approach to Homogeneous. Heterogeneous and Industrial Catalysis*, Elsevier, Amsterdam, (1993).
M. Neurock, R.A. van Santen, W. Biemolt, and A.P.J. Jansen, *J. Am. Chem. Soc.* **116**, 6860 (1994).
J.W. Niemantsverdriet, *Spectroscopy in Catalysis, an Introduction*, VCH, Weinheim, (1993).
J.W. Niemantsverdriet, A.M. van der Kraan, W.L. van Dijk, and H.S. van der Baan, *J. Phys. Chem.* **84**, 3363 (1980).
J.W. Niemantsverdriet and A.M. van der Kraan, *J. Catal.* **72**, 385 (1981).
B.E. Nieuwenhuys, *Surface Sci.* **126**, 307 (1983).
R.M. Noyes, *J. Phys. Chem.* **94**, 4404 (1990).
A. Pacault, P. Hanusse, P. de Klepper, C. Vidal, and J. Boissonade, *Acc. Chem. Res.* **9**, 438 (1976).

I. Prigogine, *From Being to Becoming*, W.H. Freeman, San Francisco, (1980).
I. Prigogine and I. Stengers, *Orde uit Chaos*, Uitgeverij Bert Bakker, Amsterdam, (1993).
P.B. Rasmussen, P.M. Holmblad, H. Christoffersen, P.A. Taylor, and I. Chorkendorff, *Surface Sci.* **287/288**, 79 (1993).
O.K. Rice, *Statistical Mechanics. Thermodynamics and Kinetics*, W.H. Freeman, San Francisco, (1967).
R.A. van Santen, *Theoretical Heterogeneous Catalysis*, World Scientific, Singapore, (1991).
R.A. van Santen and H.P.C.E. Kuipers, *Adv. Catal.* **35**, 265 (1987).
G.C.A. Schuit and L.L. van Reijen, *Adv. Catal.* **10**, 242 (1958).
T. Shimanouchi, *Tables of Molecular Vibrational Frequencies. National Standards Reference Data Series*, 39, National Bureau of Standards, Washington D.C., (1972).
E. Shustorovich, *Surface Sci. Rep.* **6**, 1 (1986).
J. Siera, Ph.D. Thesis, University of Leiden, 1992.
J.H. Sinfelt, in *Catalysis. Science and Technology* (J.R. Anderson and M. Boudart, eds.), Vol. 1, p. 257, Springer, Berlin, (1981).
G.A. Somorjai, *Chemistry in Two Dimensions*, Cornell University Press, Ithaca, (1981).
G.A. Somorjai, in *Catalyst Design, Progress and Perspectives* (L.L. Hegedus, ed.), p. 11, Wiley, New York, (1987).
G.A. Somorjai and M.A. van Hove, *Progr. Surface Sci.* **30**, 201 (1989).
J.T. Stuckless, N. Al-Sarraf, C. Wartnaby, and D.A. King, *J. Chem. Phys.* **99**, 2202 (1993).
M. Taniguchi, K. Tanaka, T. Hashizume, and T. Sakurai, *Surface Sci.* **262**, L133 (1992).
G. Tantardini, in *Cluster Models for Surface and Bulk Phenomena* (G. Pacchioni, P.S. Bagus, and F. Parmigiani, eds.), NATO ASI Series B: Physics, Vol. 283, p. 389, Plenum, New York, (1992).
K.C. Taylor, in *Catalysis and Automotive Pollution Control* (A. Crucq and A. Frennet, eds.), Elsevier, Amsterdam, (1987).
M.F.H. van Tol, Ph.D. Thesis, University of Leiden, 1993.
S. Topham, in *Catalysis. Science and Technology* (J.R. Anderson and M. Boudart, eds.), Vol. 7, p. 1, Springer, Berlin, (1985).
H. Topsøe and B.S. Clausen, *Catal. Rev.-Sci. Eng.* **26**, 395 (1984).
J. Troe, *J. Phys. Chem.* **83**, 115 (1979).
M. Vaarkamp, Ph.D. Thesis, Eindhoven University of Technology, 1993.
M.A. Vannice, *J. Catal.* **37**, 449 (1975).
M.A. Vannice, *J. Catal.* **37**, 462 (1975).
M.A. Vannice, in *Catalysis, Science and Technology* (J.R. Anderson and M. Boudart, eds.), Vol. 3, p. 139, Springer, Berlin, (1982).
W.H. Weinberg, *Langmuir* **9**, 655 (1993)
J. Weitkamp, *ACS Symposium Series* **20**, 1 (1975).
F. Zaera and G.A. Somorjai, *J. Am. Chem. Soc.* **106**, 2288 (1984).
A.N. Zaikin and A.M. Zhabotinsky, *Nature* **255**, 535 (1970).
V.P. Zhdanov, *Surface Sci. Rep.* **12**, 183 (1991).
V.P. Zhdanov, J. Pavlicek, and Z. Knor, *Catal. Rev.-Sci. Eng.* **30**, 501 (1989).

RELATED READING: GENERAL TEXTS ON KINETICS AND CATALYSIS

M. Boudart, *Kinetics of Chemical Processes*, Prentice Hall, Englewood Cliffs, (1968).
M. Boudart and G. Djéga-Mariadassou, *Kinetics of Heterogeneous Catalytic Reactions*, Princeton University Press, Princeton, (1984).
B.C. Gates, J.R. Katzer, and G.C.A. Schuit, *Chemistry of Catalytic Processes*, McGraw-Hill, New York, (1979).

G. Henrici-Olivé and S. Olivé, *Coordination and Catalysis*, VCH, Weinheirn, (1977).
K.J. Laidler, *Chemical Kinetics* (third edition), Harper and Row, New York, (1987).
R.D. Levine and R.B. Bernstein, *Molecular Reaction Dynamics*, Clarendon, Oxford, (1974).
J.A. Moulijn, P.W.N.M. van Leeuwen, and R.A. van Santen (eds.), *An Integrated Approach to Homogeneous, Heterogeneous and Industrial Catalysis*, Elsevier, Amsterdam, (1993).
J.W. Niemantsverdriet, *Spectroscopy in Catalysis: An Introduction*, VCH, Weinheim, (1993).
R.A. van Santen, *Theoretical Heterogeneous Catalysis*, World Scientific, Singapore, (1991).
J.M. Thomas and W.J. Thomas, *Introduction to Heterogeneous Catalysis*, Academic Press, New York, (1967).
W.H. Weinberg, in *Dynamics of Gas-Surface Interactions* (C.T. Rettner and M.N.R. Ashfold, eds.), The Royal Society of Chemistry, Cambridge, (1991).
V.P. Zhdanov, *Elementary Physicochemical Processes on Surfaces*, Plenum, New York, (1991).

INDEX